An Introduction to R

Notes on R: A Programming Environment for
Data Analysis and Graphics
Version 1.9.1

W. N. Venables, D. M. Smith
and the R Development Core Team

Revised and updated for publication by Network Theory Ltd.

Published by Network Theory Limited.
Third printing, April 2005. Minor corrections.
Second printing, August 2004. Revised and updated.
First printing, May 2002.

15 Royal Park
Bristol
BS8 3AL
United Kingdom

Email: info@network-theory.co.uk

ISBN 0-9541617-4-2

Original cover design by David Nicholls.

Further information about this book is available from
`http://www.network-theory.co.uk/R/intro/`

Table of Contents

Publisher's Foreword

This manual introduces the use of R, an interactive environment for statistical computing.

R is *free software*. The term "free software" refers to your freedom to run, copy, distribute, study, change and improve the software. With R you have all these freedoms.

R is part of the GNU Project. The GNU Project was launched in 1984 to develop a complete Unix-like operating system which is free software: the GNU system. It was conceived as a way of bringing back the cooperative spirit that prevailed in the computing community in earlier days, by removing the obstacles to cooperation imposed by the owners of proprietary software.

You can support the GNU Project by becoming an associate member of the Free Software Foundation and paying regular membership dues. The Free Software Foundation is a tax-exempt charity dedicated to promoting the right to use, study, copy, modify, and redistribute computer programs. It also helps to spread awareness of the ethical and political issues of freedom in the use of software. For more information, visit the website `www.fsf.org`.

The development of R itself is guided by the R Foundation, a not for profit organization working in the public interest. Individuals and organizations using R can support its continued development by becoming members of the R Foundation. Further information is available at the website `www.r-project.org`.

Brian Gough
Publisher
July 2004

Preface

This introduction to R is derived from an original set of notes describing the S and S-PLUS environments written by Bill Venables and David M. Smith (Insightful Corporation). We have made a number of small changes to reflect differences between the R and S programs, and expanded some of the material.

We would like to extend warm thanks to Bill Venables (and David Smith) for granting permission to distribute this modified version of the notes in this way, and for being a supporter of R from way back.

Comments and corrections are always welcome. Please address email correspondence to `R-core@r-project.org`.[1]

Suggestions to the reader

Most R novices will start with the introductory session in Appendix A. This should give some familiarity with the style of R sessions and more importantly some instant feedback on what actually happens.

Many users will come to R mainly for its graphical facilities. In this case, Chapter 12 [Graphics], page 93 on the graphics features can be read at almost any time and need not wait until all the preceding sections have been digested.

---

[1] Correspondence for this printed edition should be sent in the first instance to the publisher at `info@network-theory.co.uk`, as this version includes additional new material and examples. Comments will be passed onto the original authors as appropriate.

1 Introduction and preliminaries

1.1 The R environment

R is an integrated software environment for data manipulation, calculation and graphical display. Among other things it has

- an effective data handling and storage facility,
- a suite of operators for calculations on arrays, in particular matrices,
- a large, coherent, integrated collection of intermediate tools for data analysis,
- graphical facilities for data analysis and display either directly at the computer or on hardcopy, and
- a well developed, simple and effective programming language which includes conditionals, loops, user defined recursive functions and input and output facilities. (Indeed most of the system supplied functions are themselves written in this language.)

The term "environment" is intended to characterize it as a fully planned and coherent system, rather than an incremental accretion of very specific and inflexible tools, as is frequently the case with other data analysis software.

1.2 Related software and documentation

R can be regarded as an implementation of the S language which was developed at Bell Laboratories by Rick Becker, John Chambers and Allan Wilks, and also forms the basis of the S-PLUS system.

The evolution of the S language is characterized by four books by John Chambers and coauthors. For R, the basic reference is *The New S Language: A Programming Environment for Data Analysis and Graphics* by Richard A. Becker, John M. Chambers and Allan R. Wilks. The new features of the 1991 release of S (S version 3) are covered in *Statistical Models in S* edited by John M. Chambers and Trevor J. Hastie. See Appendix D [References], page 133, for precise references.

In addition, documentation for S/S-PLUS can typically be used with R, keeping the differences between the S implementations in mind. See the section "What documentation exists for R?" in the *R FAQ*, included with the R distribution[1].

[1] The R FAQ is also available from `http://www.r-project.org/`

1.3 R and statistics

Our introduction to the R environment did not mention *statistics*, yet many people use R as a statistics system.

We prefer to think of it of an environment within which many classical and modern statistical techniques have been implemented. Some of these are built into the base R environment, but many are supplied as *packages*—the distinction is largely a matter of historical accident. There are about 10 packages supplied with R (called "standard" packages) and many more are available through the Comprehensive R Archive Network (CRAN) family of Internet sites (via `http://cran.r-project.org`).

Most classical statistics and much of the latest methodology is available for use with R, but users will need to be prepared to do a little work to find it. More details on packages are given later (see Chapter 13 [Packages], page 113).

There is an important difference in philosophy between S (and hence R) and the other main statistical systems. In S, a statistical analysis is normally done as a series of steps, with intermediate results being stored in objects. Thus whereas SAS and SPSS will give copious output from a regression or discriminant analysis, R will give minimal output and store the results in a fit object for subsequent interrogation by further R functions.

1.4 R and the window system

The most convenient way to use R is at a graphics workstation running a windowing system. This guide is aimed at users who have this facility. In particular, we will occasionally refer to the use of R on an X window system, although the vast bulk of what is said applies generally to any implementation of the R environment.

Most users will find it necessary to interact directly with the operating system on their computer from time to time. In this guide, we mainly discuss interaction with the operating system on UNIX machines. If you are running R under Windows you will need to make some small adjustments.

Setting up a workstation to take full advantage of the customizable features of R is a straightforward procedure, and will not be considered further here. Users in difficulty can find further information at the R website `www.r-project.org` or seek local expert help.

1.5 Using R interactively

When you use the R program it issues a prompt when it expects input commands. The default prompt is the "greater than" sign:

```
>
```

which on UNIX might be the same as the shell prompt, and so it may appear that nothing is happening. However, as we shall see, it is easy to change to a different R prompt if you wish (see Section 10.8 [Customizing the environment], page 72). We will assume that the UNIX shell prompt is '$'.

In using R under UNIX, the suggested procedure for the first run of the program is as follows:

1. Create a separate sub-directory, say 'work', to hold data files on which you will use R for this problem, and enter the directory. This will be the working directory whenever you use R for this particular problem.

   ```
   $ mkdir work
   $ cd work
   ```

2. Start the R program with the command

   ```
   $ R
   R : Copyright 2004, The R Foundation
   Version 1.9.1 Patched (2004-07-29)
   >
   ```

3. At this point R commands may be issued (see later).
4. To quit the R program, the command is

   ```
   > q()
   Save workspace image? [y/n/c]:
   ```

   You will be asked whether you want to save the data from your R session. You can respond *yes*, *no* or *cancel* (a single letter abbreviation will do) to save the data before quitting, quit without saving, or return to the R session. Data which is saved will be available in future R sessions.

Further R sessions are simple.

1. Change to the directory 'work' and start the program as before:

   ```
   $ cd work
   $ R
   ```

2. Use the R program, terminating with the `q()` command at the end of the session.

To use R under Windows, the procedure is basically the same. Create a folder as the working directory, and set the 'Start In' field in your R shortcut to the folder path. Then launch R by double-clicking on the icon.

1.6 An introductory session

Readers wishing to get a feel for R at a computer before proceeding are strongly advised to work through the introductory session given in Appendix A [A sample session], page 117.

1.7 Getting help with functions and features

R has a built-in help facility similar to the man command of UNIX. To get more information on any specific named function, for example solve, the command is

```
> help(solve)
```

An alternative is

```
> ?solve
```

On Unix systems, the help output is displayed using the default pager. Hit the ⟨SPACE⟩ key to scroll forwards through the text, or ⟨Q⟩ to quit and return to the R prompt.

To obtain help on features specified by special characters, the argument to the help() command must be enclosed in double or single quotes, making it a "character string". For example,

```
> help("[[")
```

This is also necessary for words with syntactic meaning, including if, for and function. If in doubt, always use quotes. Either form of quote mark may be used to escape the other, as in the string "It's important". Our convention is to use double quote marks for preference.

The help.search command allows searching for entries in the help files in various ways:

```
> help.search("linear models")
Help files with alias or concept or title matching 'linear
models' using fuzzy matching:
glm.nb(MASS)        Fit a Negative Binomial Generalized
                    Linear Model
lm.gls(MASS)        Fit Linear Models by Generalized
                    Least Squares
loglm(MASS)         Fit Log-Linear Models by Iterative
                    Proportional Scaling
......              .......
```

The name of each entry is given on the left-hand side, followed by the corresponding package name in parentheses. For example, the first entry can be displayed with help(glm.nb, package="MASS"), which shows the help page for the glm.nb function in the *MASS* package. Use the command ?help.search for further details and examples.

The examples on a help topic can normally be run by

```
> example(topic)
```

On most R installations, help is also available in HTML format by running

```
> help.start()
```

which will launch a Web browser, allowing the help pages to be browsed with hyperlinks. On UNIX, subsequent help requests are sent to this HTML-based help system.

Windows versions of R have other optional help systems: use

```
> ?help
```

for further details.

1.8 R commands, case sensitivity, etc.

Technically, R is an *expression language* with a very simple syntax. It is *case sensitive*, as are most UNIX-based programs, so `A` and `a` are different symbols and refer to different variables. The set of symbols which can be used in R function and variable names depends on the operating system and its language settings (technically on the *locale* in use). Normally, all alphanumeric symbols are allowed (and in some locales this includes accented letters) plus '`.`' and '`_`', with the restriction that a name must start with '`.`' or a letter, and if it starts with '`.`' the second character must not be a digit.

Elementary commands consist of either *expressions* or *assignments*. If an expression is given as a command, it is evaluated, printed, and the value is discarded. An assignment evaluates an expression and passes the value to a variable, but the result is not automatically printed.

Commands are separated either by a semi-colon ('`;`'), or by a newline. Elementary commands can be grouped together into one compound expression by braces ('`{`' and '`}`'). *Comments* can be put almost anywhere, starting with a hashmark ('`#`'). Everything to the end of the line following a hashmark is a comment. Comments cannot be used inside strings, or in the argument list of a function definition.

If a command is not complete at the end of a line, R will give a different prompt, by default

```
+
```

on second and subsequent lines, and continue to read input until the command is syntactically complete. This prompt may be changed by the user. In this book, we will generally omit the continuation prompt and indicate continuation by simple indenting.

The result of a command is printed to the current output device. If the result is an array, such as a vector or matrix, then the elements are formatted with line breaks (if necessary) and the indices of the leading entries are labelled in square brackets, `[index]`. For example, an array of 20 elements might be displayed as follows,

```
> array(0,20)
 [1] 0 0 0 0 0 0 0 0 0 0 0 0 0 0
[15] 0 0 0 0 0 0
```

The labels '`[1]`' and '`[15]`' indicate the first and fifteenth elements in the output. It is important to note that these labels are not part of the data itself.

The index labels for matrices are placed at the start of each row and column in the output:

```
> matrix(0,4,4)
     [,1] [,2] [,3] [,4]
[1,]    0    0    0    0
[2,]    0    0    0    0
[3,]    0    0    0    0
[4,]    0    0    0    0
```

For simplicity, index labels are usually omitted in the examples in this book.

1.9 Recall and correction of previous commands

Under many versions of UNIX and on Windows, R provides a mechanism for recalling and re-executing previous commands. The vertical arrow keys on the keyboard can be used to scroll forward and backward through a *command history*. Once a command is located in this way, the cursor can be moved within the command using the horizontal arrow keys, and characters can be deleted with the ⟨DEL⟩ key or inserted with the other keys. More details are provided later: see Appendix C [The command line editor], page 129.

The recall and editing capabilities under UNIX are highly customizable. You can find out how to do this by reading the documentation for the GNU **readline** library, using `man readline` or `info readline` at the Unix prompt (if installed).

Alternatively, the GNU Emacs text editor provides more general support mechanisms (via ESS, the *Emacs Speaks Statistics* mode) for working interactively with R. See the section "R and Emacs" in the *R FAQ* for details.

1.10 Executing commands from or diverting output to a file

If commands are stored in an external file, such as 'commands.R' in the working directory 'work', they may be executed at any time in an R session with the command

```
> source("commands.R")
```

For Windows, **Source** is also available on the **File** menu. The function sink,

```
> sink("record.lis")
```

will divert all subsequent output from the console to an external file, 'record.lis'. The command

```
> sink()
```

restores output to the console once again.

1.11 Data permanency and removing objects

The entities that R creates and manipulates are known as *objects*. These may be variables, arrays of numbers, character strings, functions, or more general structures built from such components.

During an R session, objects are created and stored by name (we discuss this process in the next session). The command

```
> objects()
```

can be used to display the names of the objects which are currently stored within R. The same information can also be displayed with the ls() command. The collection of currently-stored objects is called the *workspace*.

The function rm removes objects from the workspace:

```
> rm(x, y, z, ink, junk, temp, foo, bar)
```

Objects created during an R sessions can be stored permanently in a file, for use in future R sessions. At the end of each R session you are given the opportunity to save all the currently available objects. If you indicate that you want to do this, the objects are written to a file called '.RData'(2) in the current directory.

When R is started at later time, it reloads the workspace from this file. At the same time the associated command history is reloaded.

It is recommended that you should use separate working directories for each analysis conducted with R. It is quite common for objects to be

(2) The leading "dot" in this file name makes it *invisible* in normal file listings in UNIX.

created with names such as x and y, which are meaningful in the context of a single analysis, but difficult to identify if several analyses have been carried out in the same directory.

2 Simple manipulations; numbers and vectors

2.1 Vectors and assignment

R operates on named *data structures*. The simplest such structure is the numeric *vector*, which is a single entity consisting of an ordered collection of numbers. To set up a vector named x, say, consisting of five numbers, namely 10.4, 5.6, 3.1, 6.4 and 21.7, use the R command

```
> x <- c(10.4, 5.6, 3.1, 6.4, 21.7)
```

This is an *assignment* statement using the *function* c(), which in this context can take an arbitrary number of vector *arguments*. The result is a vector obtained by concatenating the arguments end to end.(1)

A number occurring by itself in an expression is taken as a vector of length one.

Note that the assignment operator '<-' is **not** the usual '=' operator, which is reserved for another purpose. It consists of the two characters '<' ("less than") and '-' ("minus") occurring strictly side-by-side and it 'points' to the object receiving the value of the expression.

Assignments can also be made using the function assign(). An equivalent way of making the assignment shown above is

```
> assign("x", c(10.4, 5.6, 3.1, 6.4, 21.7))
```

The operator <- can be thought of as a syntactic short-cut for the assign command.

Assignments can also be made in the other direction, using '->' as an alternative assignment operator. So the same assignment could be written as

```
> c(10.4, 5.6, 3.1, 6.4, 21.7) -> x
```

If an expression is used as a complete command, the value is printed *and discarded*(2). If we were to use the command

```
> 1/x
```

the reciprocals of the five values above would be printed at the terminal (and the value of x would unchanged).

---

(1) With other than vector types of argument, such as list mode arguments, the action of c() is rather different. See Section 6.2.1 [Concatenating lists], page 42.

(2) Actually, the previous value is still available as .Last.value before any other statements are executed

The further assignment

```
> y <- c(x, 0, x)
```

would create a vector y with 11 entries, consisting of two copies of x with a zero in the middle place.

2.2 Vector arithmetic

Vectors can be used in arithmetic expressions, in which case the operations are performed element-by-element. Vectors occurring in the same expression need not all be of the same length. If they are not, the resulting value of the expression is a vector with the same length as the longest vector which occurs in the expression. Shorter vectors in the expression are *recycled* as often as need be (perhaps fractionally) until they match the length of the longest vector. In particular, a constant is simply repeated. So with the above assignments the command

```
> v <- 2*x + y + 1
```

generates a new vector v of length 11 constructed by adding together, element-by-element, 2*x repeated 2.2 times, y repeated just once, and 1 repeated 11 times.

The elementary arithmetic operators use the normal symbols +, -, *, /, and the exponentiation operator ^ for raising to a power. In addition, all of the common mathematical functions are available: log, exp, sin, cos, tan, sqrt, and so on, with their usual meanings. The functions max and min select the largest and smallest elements of a vector respectively. range is a function whose value is a vector of length two, namely c(min(x), max(x)). The function length(x) is the number of elements in x, sum(x) gives the total of the elements in x, and prod(x) their product.

Two statistical functions are mean(x) which calculates the sample mean,

```
sum(x)/length(x)
```

and var(x) which gives the sample variance,

```
sum((x-mean(x))^2)/(length(x)-1)
```

If the argument to var() is an n-by-p matrix, the result is a p-by-p sample covariance matrix found by regarding the rows as independent p-variate sample vectors.

sort(x) returns a vector of the same size as x with the elements arranged in increasing order. There are also other more flexible sorting functions available, such as order() or sort.list(), which produce a permutation to do the sorting.

Note that max and min select the largest and smallest values in their arguments, even if they are given several vectors. The *parallel* maximum and minimum functions `pmax` and `pmin` return a vector (of length equal to their longest argument) that contains in each element the maximum or minimum element in that position in any of the input vectors.

For most purposes, the user will not be concerned if the "numbers" in a numeric vector are integers, reals or even complex. Internally, calculations are done as double precision real numbers, or double precision complex numbers if the input data are complex.

To work with complex numbers, supply an explicit complex part. Thus, the expression

```
sqrt(-17)
```

will give `NaN` (*Not a Number*) and a warning, but

```
sqrt(-17+0i)
```

will do the computations as complex numbers.

2.3 Generating regular sequences

R has a number of commands for generating sequences of numbers.

The most important sequence operator is the colon ':', which produces a linear range of values. For example, `1:30` is the vector

```
c(1, 2, 3, ..., 29, 30)
```

The colon operator has highest priority within an expression, so that `2*1:15` is equivalent to `2*(1:15)`, giving the vector

```
c(2, 4, 6, ..., 28, 30)
```

Put `n <- 10` and compare the sequences `1:n-1` and `1:(n-1)` to see the difference. For decreasing sequences, the construction `n:1` may be used to generate a sequence backwards.

The function `seq()` is a more general function for generating sequences. It has five arguments, only some of which may be specified in any one call. The first two arguments, if given, specify the beginning and end of the sequence. If these are the only two arguments given the result is the same as the colon operator, i.e. `seq(2,10)` is equivalent to `2:10`.

Parameters to `seq()`, and to many other R functions, can also be given in named form, in which case the order in which they appear is irrelevant. The first two parameters may be named `from=`*value* and `to=`*value*; thus `seq(1,30)`, `seq(from=1, to=30)` and `seq(to=30, from=1)` are all the same as `1:30`. The next two parameters to `seq()` may be named `by=`*value* and `length=`*value*, which specify a step size and a length for the sequence respectively. If neither of these is given, the default `by=1` is assumed.

For example,

```
> seq(-5, 5, by=.2) -> s3
```

generates the vector `c(-5.0, -4.8, -4.6, ..., 4.6, 4.8, 5.0)` and stores it in `s3`. Similarly,

```
> s4 <- seq(length=51, from=-5, by=.2)
```

generates the same vector in `s4`.

The fifth parameter may be named `along=`*`vector`*. If used, `along=` must be the only parameter, and creates a sequence `1, 2, ...,` `length(`*`vector`*`)`, or the empty sequence if the vector is empty (as it can be).

A related function is `rep()` which replicates objects in various ways. The simplest form is

```
> s5 <- rep(x, times=5)
```

which will put five copies of `x` end-to-end in `s5`.

2.4 Logical vectors

As well as numerical vectors, R also allows manipulation of logical quantities. The elements of a logical vector can have the values `TRUE`, `FALSE`, and `NA` (for "not available", see below). The first two are often abbreviated as `T` and `F`, respectively. Note however that `T` and `F` are just variables which are set to `TRUE` and `FALSE` by default, but are not reserved words and can be overwritten by the user. Hence, you should always use `TRUE` and `FALSE`.

Logical vectors are generated by *conditions*. For example,

```
> temp <- x > 13
```

sets `temp` as a vector of the same length as `x` with values `FALSE` corresponding to elements of `x` where the condition is *not* met and `TRUE` where it is.

The logical operators are `<`, `<=`, `>`, `>=`, `==` for exact equality and `!=` for inequality. In addition, if `c1` and `c2` are logical expressions, then `c1 & c2` is their intersection (*"and"*), `c1 | c2` is their union (*"or"*), and `!c1` is the negation of `c1`.

Logical vectors may be used in ordinary arithmetic, where they are *coerced* into numeric vectors, `FALSE` becoming `0` and `TRUE` becoming `1`. However, there are situations where logical vectors and their coerced numeric counterparts are not equivalent, as in the following examples.

2.5 Missing values

In some cases, the components of a vector may not be completely known. When an element or value is "not available" or a "missing value" (in the statistical sense), a place within a vector may be reserved for it by assigning it the special value `NA`. In general, any operation on an `NA` becomes an `NA`. The motivation for this rule is simply that if the specification of an operation is incomplete the result cannot be known, and hence is not available.

The function `is.na(x)` gives a logical vector of the same size as `x` with value `TRUE` if and only if the corresponding element in `x` is `NA`.

```
> z <- c(1:3,NA);  ind <- is.na(z)
```

Notice that the logical expression `x == NA` is quite different from `is.na(x)` since `NA` is not really a value but a marker for a quantity that is not available. Thus `x == NA` is a vector of the same length as `x` *all* of whose values are `NA`, since the logical expression itself is incomplete and hence undecidable.

Note that there is a second kind of "missing" value produced by numerical computation, the so-called *Not a Number*, `NaN`, value. Examples are

```
> 0/0
```

or

```
> Inf - Inf
```

which both give `NaN` since the result cannot be defined sensibly. The presence of `NaN` values can be tested using the function `is.nan(x)`.

In summary, the function `is.na(x)` is `TRUE` *both* for `NA` and `NaN` values, while `is.nan(x)` is only `TRUE` for `NaN`s.

2.6 Character vectors

Character quantities and character vectors are used frequently in R, for example as plot labels and titles. They are denoted by a sequence of characters delimited by the double quote character, e.g., `"x-values"`, `"New iteration results"`.

Character strings can be entered using either double (`"`) or single (`'`) quotes, but are printed using double quotes (or sometimes without quotes). They use C-style escape sequences with backslash (`\`) as the escape character. The backslash character itself is entered and printed as `\\` inside double quotes, while the double-quote character (`"`) can be entered as `\"`. Other useful escape sequences are `\n`, newline, `\t`, tab and `\b`, backspace.

Character vectors may be concatenated into a vector by the `c()` function.

The `paste()` function takes an arbitrary number of arguments and concatenates them one by one into character strings. Any numbers given among the arguments are coerced into character strings, in the same way they would be if they were printed. By default, the arguments are separated by a single blank character in the result, but this can be controlled with the named parameter `sep=string`, which changes the separator to *string* (which may also be an empty string).

For example,

```
> labs <- paste(c("X","Y"), 1:10, sep="")
```

makes `labs` into the character vector

```
c("X1", "Y2", "X3", "Y4", "X5", "Y6", "X7", "Y8",
  "X9", "Y10")
```

The function `paste(..., collapse=ss)` joins its arguments into a single character string, putting *ss* in between. Note that recycling of short lists takes place here too; thus `c("X", "Y")` is repeated 5 times to match the sequence `1:10`. There are more tools for character manipulation, see the entries for `sub` and `substring` in the R Reference Manual or help pages for details.

2.7 Index vectors; selecting and modifying subsets of a data set

Subsets of the elements of a vector may be selected by appending to the name of the vector an *index vector* in square brackets, `v[i]`. More generally, any expression that evaluates to a vector may have subsets of its elements similarly selected by appending an index vector in square brackets immediately after the expression, `(expression)[i]`.

Such index vectors can be any of four distinct types.

1. **A logical vector**. In this case, the index vector must be of the same length as the vector from which elements are to be selected. Values corresponding to `TRUE` in the index vector are selected and those corresponding to `FALSE` are omitted. For example,

   ```
   > y <- x[!is.na(x)]
   ```

   creates (or re-creates) an object `y` which will contain the non-missing values of `x`, in the same order. Note that if `x` has missing values, `y` will be shorter than `x`. Also

   ```
   > (x+1)[(!is.na(x)) & x>0] -> z
   ```

   creates an object `z` and places in it the values of the vector `x+1` for which the corresponding value in `x` was both non-missing and positive.

2. **A vector of positive integral quantities**. In this case, the values in the index vector must lie in the set {1, 2, ..., `length(x)`}. The corresponding elements of the vector are selected and concatenated,

in that order, in the result. The index vector can be of any length and the result is of the same length as the index vector. For example, `x[6]` is the sixth component of `x` and

```
> x[1:10]
```

selects the first 10 elements of `x` (assuming `length(x)` is not less than 10). Also

```
> c("x","y")[rep(c(1,2,2,1), times=4)]
```

(an admittedly unlikely thing to do) produces a character vector of length 16 consisting of `"x"`, `"y"`, `"y"`, `"x"` repeated four times.

3. **A vector of negative integral quantities**. Such an index vector specifies the values to be *excluded* rather than included. Thus

```
> y <- x[-(1:5)]
```

gives `y` all but the first five elements of `x`.

4. **A vector of character strings**. This possibility only applies where an object has a names attribute to identify its components. In this case, a sub-vector of the names vector may be used in the same way as the positive integral labels in item 2 above.

```
> fruit <- c(5, 10, 1, 20)
> names(fruit) <- c("orange", "banana", "apple",
                    "peach")
> lunch <- fruit[c("apple","orange")]
```

The advantage of alphanumeric *names* is that they are often easier to remember than *numeric indices.* This option is particularly useful in connection with data frames, as we shall see later.

An indexed expression can also appear on the receiving end of an assignment, in which case the assignment operation is performed *only on those elements of the vector.* The expression must be of the form `vector[index_vector]` as having an arbitrary expression in place of the vector name does not make much sense here.

The vector assigned must match the length of the index vector, and in the case of a logical index vector it must again be the same length as the vector it is indexing.

For example,

```
> x[is.na(x)] <- 0
```

replaces any missing values in `x` by zeros and

```
> y[y < 0] <- -y[y < 0]
```

has the same effect as

```
> y <- abs(y)
```

2.8 Other types of objects

Vectors are the most important type of object in R, but there are several others which we will meet more formally in later sections.

- *matrices* or more generally *arrays* are multidimensional generalizations of vectors. In fact, they *are* vectors that can be indexed by two or more indices and will be printed in special ways. See Chapter 5 [Arrays and matrices], page 29.
- *factors* provide compact ways to handle categorical data. See Chapter 4 [Factors], page 25.
- *lists* are a general form of vector in which the various elements need not be of the same type, and are often themselves vectors or lists. Lists provide a convenient way to return the results of a statistical computation. See Section 6.1 [Lists], page 41.
- *data frames* are matrix-like structures, in which the columns can be of different types. Think of data frames as 'data matrices' with one row per observational unit but with (possibly) both numerical and categorical variables. Many experiments are best described by data frames: the treatments are categorical but the response is numeric. See Section 6.3 [Data frames], page 42.
- *functions* are themselves objects in R which can be stored in the project's workspace. This provides a simple and convenient way to extend R. See Chapter 10 [Writing your own functions], page 63.

3 Objects, their modes and attributes

3.1 Intrinsic attributes: mode and length

The entities R operates on are technically known as *objects*. Examples are vectors of numeric (real) or complex values, vectors of logical values and vectors of character strings. These are known as "atomic" structures since their components are all of the same type, or *mode*, namely *numeric*[1], *complex*, *logical* and *character* respectively.

Vectors must have their values *all of the same mode*. Thus any given vector must be unambiguously either *logical*, *numeric*, *complex* or *character*. The only mild exception to this rule is the special "value" listed as `NA` for quantities not available. Note that a vector can be empty and still have a mode. For example, the empty character string vector is listed as `character(0)` and the empty numeric vector as `numeric(0)`.

R also operates on objects called *lists*, which are of mode *list*. These are ordered sequences of objects which individually can be of any mode. Lists are known as "recursive" rather than atomic structures, since their components can themselves be lists in their own right.

The other recursive structures are those of mode *function* and *expression*. Functions are the objects that form part of the R system along with similar user written functions, which we discuss in some detail later. Expressions as objects form an advanced part of R which will not be covered in this introduction, except indirectly when we discuss *formulae* used with modeling in R.

By the *mode* of an object we mean the basic type of its fundamental constituents. This is a special case of a "property" of an object. Another property of every object is its *length*. The functions `mode(object)` and `length(object)` can be used to find out the mode and length of any defined structure[2].

Further properties of an object can be found using the function `attributes(object)`, see Section 3.3 [Getting and setting attributes], page 23. Because of this, *mode* and *length* are also called "intrinsic attributes" of an object.

---

[1] *numeric* mode is actually an amalgam of two distinct modes, namely *integer* and *double* precision, as explained in the R Reference Manual.

[2] Note however that `length(object)` does not always contain intrinsic useful information, e.g., when `object` is a function.

For example, if z is a complex vector of length 100, then in an expression mode(z) is the character string "complex" and length(z) is 100.

R caters for changes of mode almost anywhere it could be considered sensible to do so (and a few where it might not be). For example, with

```
> z <- 0:9
```

we could put

```
> digits <- as.character(z)
```

after which digits is the character vector c("0", "1", "2", ..., "9"). A further *coercion*, or change of mode, reconstructs the numerical vector again:

```
> d <- as.integer(digits)
```

Now d and z are the same.[3] There is a large collection of functions of the form as.*something*() for either coercion from one mode to another, or for investing an object with some other attribute it may not already possess. The reader should consult the R Reference Manual or help pages to become familiar with them.

3.2 Changing the length of an object

An "empty" object may still have a mode. For example,

```
> e <- numeric()
```

makes e an empty vector structure of mode numeric. Similarly, the expression character() is a empty character vector, and so on. Once an object of any size has been created, new components may be added to it simply by giving it an index value outside its previous range. Thus

```
> e[3] <- 17
```

now makes e a vector of length 3 (the first two components of which are at this point both NA). This applies to any structure at all, provided the mode of the additional component(s) agrees with the mode of the object in the first place.

This automatic adjustment of lengths of an object is used often, for example in the scan() function for input (see Section 7.2 [The scan() function], page 48).

Conversely, to truncate the size of an object requires only an assignment to do so. Hence if alpha is an object of length 10, then

```
> alpha <- alpha[2 * 1:5]
```

makes it an object of length 5 consisting of just the former components with even index. The old indices are not retained, of course.

---

[3] In general, coercion from numeric to character and back again will not be exactly reversible, because of roundoff errors in the character representation.

3.3 Getting and setting attributes

The function `attributes(object)` gives a list of all the non-intrinsic attributes currently defined for that object. The function `attr(object, name)` can be used to select a specific attribute. These functions are rarely used, except in rather special circumstances when some new attribute is being created for some particular purpose, for example to associate a creation date or an operator with an R object. The concept, however, is very important.

Some care should be exercised when assigning or deleting attributes since they are an integral part of the object system used in R.

When the `attribute` function is used on the left hand side of an assignment, it can either associate a new attribute with `object` or change an existing one. For example,

```
> attr(z,"dim") <- c(10,10)
```

allows R to treat `z` as if it were a 10-by-10 matrix.

3.4 The class of an object

A special attribute known as the *class* of the object allows an object-oriented style of programming in R.

For example, if an object has class `"data.frame"`, it will be printed in a certain way specific to this class, the `plot()` function will display it graphically according to its class, and other so-called generic functions such as `summary()` will react to it as an argument in a way sensitive to its class.

To remove temporarily the effects of class, use the function `unclass()`. For example, if `winter` has the class `"data.frame"` then

```
> winter
```

will print it in data frame form, which is rather like a matrix, whereas

```
> unclass(winter)
```

will print it as an ordinary list. Only in rather special situations do you need to use this, but one is when you are learning to come to terms with the idea of class and generic functions.

Generic functions and classes will be discussed further in Section 10.9 [Object orientation], page 73, but only briefly.

4 Ordered and unordered factors

A *factor* is a vector object used to specify a discrete classification (grouping) of the components of other vectors of the same length. R provides both *ordered* and *unordered* factors. While the "real" application of factors is with model formulae (see Section 11.1.1 [Contrasts], page 78), here we look at a specific example.

4.1 A specific example

Suppose, for example, we have a sample of 30 tax accountants from all the states and territories of Australia[1] and their individual state of origin is specified by a character vector of state mnemonics as

```
> state <- c("tas", "sa",  "qld", "nsw", "nsw", "nt",
             "wa",  "wa",  "qld", "vic", "nsw", "vic",
             "qld", "qld", "sa",  "tas", "sa",  "nt",
             "wa",  "vic", "qld", "nsw", "nsw", "wa",
             "sa",  "act", "nsw", "vic", "vic", "act")
```

This vector can be *encoded* using the different element values, allowing other R commands to operate on these subsets. The encoded form of the vector, known as a *factor*, is created using the `factor()` function:

```
> statef <- factor(state)
```

and displayed like this:

```
> statef
 [1] tas sa  qld nsw nsw nt  wa  wa
     qld vic nsw vic qld qld sa tas
[16] sa  nt  wa  vic qld nsw nsw wa
     sa  act nsw vic vic act
Levels:  act nsw nt qld sa tas vic wa
```

The discrete categories of the factor are referred to as *levels* (in this case, the levels are the abbreviated names of the states). To find out the levels of a factor, the function `levels()` can be used:

```
> levels(statef)
"act" "nsw" "nt" "qld" "sa" "tas" "vic" "wa"
```

---

[1] Foreign readers should note that there are eight states and territories in Australia, namely the Australian Capital Territory, New South Wales, the Northern Territory, Queensland, South Australia, Tasmania, Victoria and Western Australia.

Note that in the case of character vectors, the levels are sorted in alphabetical order.

4.2 The function tapply() and ragged arrays

To continue the previous example, suppose we have the incomes of the same tax accountants in another vector (in suitably large units of money)

```
> incomes <- c(60, 49, 40, 61, 64, 60, 59, 54, 62, 69,
               70, 42, 56, 61, 61, 61, 58, 51, 48, 65,
               49, 49, 41, 48, 52, 46, 59, 46, 58, 43)
```

To calculate the sample mean income for each state we can now use the special function `tapply()`,

```
> incmeans <- tapply(incomes, statef, mean)
```

This gives a means vector with the components labelled by the levels:

```
   act    nsw     nt    qld     sa    tas    vic     wa
44.500 57.333 55.500 53.600 55.000 60.500 56.000 52.250
```

The function `tapply()` is used to apply a function, here `mean()`, to each group of components of its first argument, here `incomes`, defined by the levels of the second component, here `statef`[2], as if they were separate vector structures. The result is a structure of the same length as the levels attribute of the factor containing the results. The reader should consult the R Reference Manual or the help pages for more details about the `tapply` command.

Suppose further we needed to calculate the standard errors of the state income means. To do this we need to write an R function to calculate the standard error for any given vector. Since there is a builtin function `var()` to calculate the sample variance, such a function is a very simple one liner, specified by the assignment:

```
> stderr <- function(x) sqrt(var(x)/length(x))
```

(Writing functions will be considered later in Chapter 10 [Writing your own functions], page 63). After this assignment, the standard errors are calculated by

```
> incster <- tapply(incomes, statef, stderr)
```

and the values calculated are then

```
> incster
act    nsw  nt    qld     sa tas   vic     wa
1.5 4.3102 4.5 4.1061 2.7386 0.5 5.244 2.6575
```

---

[2] Note that `tapply()` also works in this case when its second argument is not a factor, e.g., '`tapply(incomes, state)`', and this is true for quite a few other functions, since arguments are *coerced* to factors when necessary (using `as.factor()`).

As an exercise you may care to find the usual 95% confidence limits for the state mean incomes. To do this you could use `tapply()` once more with the `length()` function to find the sample sizes, and the `qt()` function to find the percentage points of the appropriate t-distributions.

The function `tapply()` can be used to handle more complicated indexing of a vector by multiple categories. For example, we might wish to split the tax accountants by both state and sex. However in this simple instance (just one category) what happens can be thought of as follows. The values in the vector are collected into groups corresponding to the distinct entries in the category. The function is then applied to each of these groups individually. The value is a vector of function results, labelled by the levels attribute of the category.

The combination of a vector and a labelling factor is an example of what is sometimes called a *ragged array*, since the subclass sizes are possibly irregular. When the subclass sizes are all the same the indexing may be done implicitly, and much more efficiently, as we see in the next section.

4.3 Ordered factors

The levels of factors are stored in alphabetical order, or in the order they were specified to `factor` if they were specified explicitly.

Sometimes the levels will have a natural ordering that we want to record and want our statistical analysis to make use of. The `ordered()` function creates such ordered factors, but is otherwise identical to `factor`. For most purposes, the only difference between ordered and unordered factors is that the former are printed showing the ordering of the levels, but the contrasts generated for them in fitting linear models are also different.

5 Arrays and matrices

5.1 Arrays

An array can be considered as a multiply subscripted collection of data entries, such as floating-point numbers, integers or logical values. R provides simple facilities for creating and handling multidimensional arrays, and in particular the special case of matrices (2-dimensional arrays).

Every array has a dimension vector, which is a vector of positive integers. If the length of the dimension vector is k then the array is k-dimensional. The values in the dimension vector give the upper limits for each of the k subscripts. The lower limits are always 1.

A vector can be used by R as an array only if it has a dimension vector as its *dim* attribute. Suppose, for example, z is a vector of 1500 elements. The assignment

```
> dim(z) <- c(3,5,100)
```

gives it the *dim* attribute that allows it to be treated as a 3 by 5 by 100 array.

Other functions such as `matrix()` and `array()` are available for simpler and more natural looking assignments, as we shall see in Section 5.4 [The array() function], page 31.

The values in the data vector give the values in the array in the same order as they would occur in FORTRAN, that is "column major order," with the first subscript moving fastest and the last subscript slowest.

For example, if the dimension vector for an array a is `c(3,4,2)` then there are $3 * 4 * 2 = 24$ entries in a, and the data vector holds them in the order `a[1,1,1]`, `a[2,1,1]`, ..., `a[2,4,2]`, `a[3,4,2]`.

5.2 Array indexing. Subsections of an array

Individual elements of an array may be referenced, as above, by giving the name of the array followed by the subscripts in square brackets, separated by commas.

More generally, subsections of an array may be specified by giving a sequence of *index vectors* in place of subscripts; however *if any index position is given an empty index vector, then the full range of that subscript is taken.*

Continuing the previous example, `a[2,,]` is a $4 * 2$ array with dimension vector `c(4,2)` and data vector containing the values

```
c(a[2,1,1], a[2,2,1], a[2,3,1], a[2,4,1],
  a[2,1,2], a[2,2,2], a[2,3,2], a[2,4,2])
```

in that order. `a[,,]` stands for the entire array, which is the same as omitting the subscripts entirely and using `a` alone.

For any array, say `Z`, the dimension vector may be referenced explicitly as `dim(Z)` (on either side of an assignment).

Also, if an array name is given with just *one subscript or index vector*, then the corresponding values of the data vector only are used; in this case the dimension vector is ignored. This is not the case, however, if the single index is not a vector but itself an array, as we next discuss.

5.3 Index arrays

As well as an index vector in any subscript position, an array may be used with a single *index array* in order to assign a vector of quantities to an irregular collection of elements in the array, or to extract an irregular collection as a vector.

The following matrix example should make the process clear. In the case of a doubly indexed array, an index matrix may be given consisting of two columns and as many rows as desired. The entries in the index matrix are the row and column indices for the doubly indexed array. Suppose we have a 4 by 5 array `X` and we wish to do the following:

- Extract elements `X[1,3]`, `X[2,2]` and `X[3,1]` as a vector structure, and
- Replace these entries in the array `X` by zero.

In this case we need a 3 by 2 subscript array, as in the following example:

```
> x <- array(1:20,dim=c(4,5))    # Generate a 4 by 5 array.
> x
     [,1] [,2] [,3] [,4] [,5]
[1,]    1    5    9   13   17
[2,]    2    6   10   14   18
[3,]    3    7   11   15   19
[4,]    4    8   12   16   20
> i <- array(c(1:3,3:1),dim=c(3,2))
> i                              # i is a 3 by 2 index array.
     [,1] [,2]
[1,]    1    3
[2,]    2    2
[3,]    3    1
> x[i]                           # Extract those elements
9 6 3
```

```
> x[i] <- 0                    # Replace those elements by zeros.
> x
     [,1] [,2] [,3] [,4] [,5]
[1,]    1    5    0   13   17
[2,]    2    0   10   14   18
[3,]    0    7   11   15   19
[4,]    4    8   12   16   20
>
```

As a less trivial example, suppose we wish to generate an unreduced design matrix for a block design defined by factors `blocks` (b levels) and `varieties` (v levels). Further suppose there are n plots in the experiment. We could proceed as follows:

```
> Xb <- matrix(0, n, b)
> Xv <- matrix(0, n, v)
> ib <- cbind(1:n, blocks)
> iv <- cbind(1:n, varieties)
> Xb[ib] <- 1
> Xv[iv] <- 1
> X <- cbind(Xb, Xv)
```

To construct the incidence matrix, `N`, we could use

```
> N <- crossprod(Xb, Xv)
```

However, a simpler direct way of producing this matrix is to use the function `table()`:

```
> N <- table(blocks, varieties)
```

5.4 The `array()` function

As well as giving a vector structure a `dim` attribute, arrays can be constructed from vectors by the `array` function, which has the form

```
> Z <- array(data_vector, dim_vector)
```

For example, if the vector `h` contains 24, or fewer, numbers then the command

```
> Z <- array(h, dim=c(3,4,2))
```

would use `h` to set up 3 by 4 by 2 array in `Z`. If the size of `h` is exactly 24 the result is the same as

```
> dim(Z) <- c(3,4,2)
```

However, if `h` is shorter than 24 then its values are recycled from the beginning again, to make it up to size 24 (see Section 5.4.1 [The recycling rule], page 32). As an extreme but common example

```
> Z <- array(0, c(3,4,2))
```

makes `Z` an array of all zeros.

At this point dim(Z) stands for the dimension vector c(3,4,2), and Z[1:24] stands for the data vector as it was in h, and Z[] with an empty subscript or Z with no subscript stands for the entire array as an array.

Arrays may be used in arithmetic expressions and the result is an array formed by element-by-element operations on the data vector. The dim attributes of operands generally need to be the same, and this becomes the dimension vector of the result. So if A, B and C are all similar arrays, then

```
> D <- 2*A*B + C + 1
```

makes D a similar array, with its data vector being the result of the given element-by-element operations. However, if the arrays have different sizes then the precise rules for mixed array and vector calculations have to be considered a little more carefully.

5.4.1 Mixed vector and array arithmetic. The recycling rule

The rules affecting element-by-element mixed calculations with vectors and arrays are somewhat complicated. From experience, we have found the following to be a reliable guide:

- The expression is scanned from left to right.
- Any short vector operands are extended by recycling their values until they match the size of any other operands.
- If short vectors and arrays *only* are encountered, the arrays must all have the same dim attribute or an error results.
- Any vector operand longer than a matrix or array operand generates an error.
- If array structures are present and no error or coercion to vector has been precipitated, the result is an array structure with the common dim attribute of its array operands.

5.5 The outer product of two arrays

An important operation on arrays is the *outer product*. If a and b are two numeric arrays then their outer product is an array whose dimension vector is given by concatenating their two dimension vectors in order, and whose data vector is found by forming all possible products of the elements of the data vector of a with those of b. The outer product is computed using the special operator %o%:

```
> ab <- a %o% b
```

An alternative form is

```
> ab <- outer(a, b, "*")
```

When using the outer() function, the multiplication operator '*' can be replaced by an arbitrary function of two variables. For example, if we wished to evaluate the function $f(x;y) = \cos(y)/(1+x^2)$ over a regular grid of points with x- and y-coordinates defined by the R vectors a and b respectively, we could proceed as follows:

```
> f <- function(x, y) cos(y)/(1 + x^2)
> z <- outer(a, b, f)
```

The outer product of two ordinary vectors is a matrix (i.e. a doubly subscripted array). Note that the outer product operator is non-commutative. Defining your own R functions will be considered further in Chapter 10 [Writing your own functions], page 63.

An example: Determinants of 2 by 2 digit matrices

As an artificial but cute example, consider the determinants of 2 by 2 matrices $[a, b; c, d]$ where each entry is a non-negative integer in the range $0, 1, \ldots, 9$, that is a digit.

The problem is to find the determinants, $ad-bc$, of all possible matrices of this form and represent the frequency with which each value occurs as a *high density* plot. This amounts to finding the probability distribution of the determinant if each digit is chosen independently and uniformly at random.

A neat way of doing this uses the outer() function twice:

```
> d <- outer(0:9, 0:9)
> fr <- table(outer(d, d, "-"))
> plot(as.numeric(names(fr)), fr, type="h",
       xlab="Determinant", ylab="Frequency")
```

Notice the coercion of the names attribute of the frequency table to numeric, in order to recover the range of the determinant values. The "obvious" way of doing this problem with for loops, to be discussed in Chapter 9 [Loops and conditional execution], page 61, is so inefficient as to be impractical.

It is also perhaps surprising that about 1 in 20 such matrices is singular.

5.6 Generalized transpose of an array

The function aperm(a, perm) may be used to permute the dimensions of an array, a. The argument perm must be a permutation of the integers $\{1, \ldots, k\}$, where k is the number of subscripts in a. The result of the function is an array of the same size as a, but with the old dimension given by perm[j] becoming the new j-th dimension. The easiest way to think of this operation is as a generalization of transposition for matrices. Indeed, if A is a matrix, (that is, a doubly subscripted array) then the matrix B, given by

```
> B <- aperm(A, c(2,1))
```

is just the transpose of A. For this special case a simpler function t() is available, so we could have used B <- t(A).

5.7 Matrix facilities

As noted above, a matrix is just an array with two subscripts. However, it is such an important special case it needs a separate discussion. R contains many operators and functions that are available only for matrices. For example, t(X) is the matrix transpose function, and only works with two-dimensional arrays. The functions nrow(A) and ncol(A) give the number of rows and columns in the matrix A respectively.

5.7.1 Matrix multiplication

The operator %*% is used to form the product of two matrices. For example, if A and B are square matrices of the same size then

```
> C <- A %*% B
```

is the matrix multiplication $C = AB$, while

```
> C <- A * B
```

is the matrix of element-by-element products, $C_{ij} = A_{ij} * B_{ij}$.

If x is a vector, then

```
> x %*% A %*% x
```

is a quadratic form.[1]

An n-by-1 or 1-by-n matrix may be used as an n-vector, in the appropriate context. Similarly, vectors which occur in matrix multiplication expressions are automatically promoted to row or column vectors, where possible, whichever is multiplicatively conformant. This is not always unambiguously possible, as we see later.

The function crossprod() forms "cross products", meaning that crossprod(X, y) is the same as t(X) %*% y but the operation is more

(1) Note that x %*% x is ambiguous, as it could mean either x'x or xx', where x is the column form. In such cases, the smaller matrix interpretation is adopted, so the scalar x'x is the result. The matrix xx' may be calculated unambiguously either by cbind(x) %*% x or x %*% rbind(x) since the result of rbind() or cbind() is always a matrix.

efficient. If the second argument to `crossprod()` is omitted it is taken to be the same as the first.

Diagonal matrices can be created with the `diag()` function in several different ways—the meaning of `diag()` depends on its argument. When the argument is a vector, `diag(v)` gives a diagonal matrix with the elements of the vector as the diagonal entries. On the other hand, `diag(M)` gives the vector of main diagonal entries of `M`, when `M` is a matrix. This is the same convention as that used for `diag()` in MATLAB. Also, if `k` is a single numeric value then `diag(k)` is the `k`-by-`k` identity matrix.

5.7.2 Linear equations and inversion

Solving linear equations is the inverse of matrix multiplication. When after

```
> b <- A %*% x
```

only `A` and `b` are given, the vector `x` is the solution of that linear equation system. In R,

```
> solve(A,b)
```

solves the system $Ax = b$, returning `x` (up to some accuracy loss). Note that in linear algebra, formally $\mathbf{x} = \mathbf{A}^{-1}\mathbf{b}$ where $\mathbf{A}^{-1}$ denotes the *inverse* of $\mathbf{A}$, which can be computed by

```
solve(A)
```

but rarely is needed. Numerically, it is both more efficient and more stable to compute `solve(A,b)` instead of `x <- solve(A) %*% b`.

Similarly, the quadratic form $\mathbf{x}'\mathbf{A}^{-1}\mathbf{x}$, found in multivariate systems, should be computed using an expression like `x %*% solve(A,x)`, rather than computing the inverse of `A`.

5.7.3 Eigenvalues and eigenvectors

The function `eigen(S)` calculates the eigenvalues and eigenvectors of a symmetric matrix `S`. The result of this function is a list of two components named `values` and `vectors`. The assignment

```
> ev <- eigen(S)
```

will assign this list to `ev`. Then `ev$val` is the vector of eigenvalues of `S` and `ev$vec` is the matrix of corresponding eigenvectors. Had we only needed the eigenvalues, we could have used the assignment:

```
> evals <- eigen(S)$values
```

`evals` now holds the vector of eigenvalues and the second component is discarded. If the expression

```
> eigen(S)
```

is used by itself as a command the two components are printed, with their names.

5.7.4 Singular value decomposition and determinants

The function svd(M) takes an arbitrary matrix argument, M, and calculates its singular value decomposition (SVD). The SVD consists of a matrix of orthonormal columns U with the same column space as M, a second matrix of orthonormal columns V, whose column space is the row space of M, and a diagonal matrix of positive entries D such that M = U %*% D %*% t(V), where D is returned as a vector of the diagonal elements. The result of svd(M) is a list of three components named d, u and v.

If M is square then the absolute value of the determinant of M can be computed with

```
> absdetM <- prod(svd(M)$d)
```

If this calculation were needed often with a variety of matrices, it could be defined as an R function

```
> absdet <- function(M) prod(svd(M)$d)
```

after which we could use absdet() as just another R function. As a further trivial but potentially useful example, you might like to consider writing a function, say tr(), to calculate the trace of a square matrix. [Hint: You will not need to use an explicit loop. Look again at the diag() function.]

5.7.5 Least squares fitting and the QR decomposition

The function lsfit() performs a least-squares fitting procedure, returning a list of results, such as the best-fit coefficients and residuals. An assignment such as

```
> ans <- lsfit(X, y)
```

gives the results of a least squares fit, where y is the vector of observations and X is the design matrix. See the R Reference Manual or help pages for more details of the lsfit() command, and also for the follow-up function ls.diag() for regression diagnostics. Note that a grand mean term is automatically included and need not be included explicitly as a column of X. Further note that you almost always will prefer using lm(.) (see Section 11.2 [Linear models], page 79) to lsfit() for regression modeling.

Another closely related function is qr() and its allies. Consider the following assignments

```
> Xplus <- qr(X)
```

```
> b <- qr.coef(Xplus, y)
> fit <- qr.fitted(Xplus, y)
> res <- qr.resid(Xplus, y)
```

These compute the orthogonal projection of y onto the range of X in fit, the projection onto the orthogonal complement in res and the coefficient vector for the projection in b, that is, b is essentially the result of the MATLAB 'backslash' operator.

When using the qr() decomposition, it is not assumed that X has full column rank—any redundancies will be discovered and removed as they are found.

This alternative is the older, low-level way to perform least squares calculations. Although still useful in some contexts, it would now generally be replaced by the statistical models features, as will be discussed in Chapter 11 [Statistical models in R], page 75.

5.8 Forming partitioned matrices, cbind() and rbind()

As we have already seen informally, matrices can be built up from other vectors and matrices by the functions cbind() and rbind(). Roughly speaking, cbind() forms matrices by binding together matrices horizontally, or column-wise, and rbind() vertically, or row-wise.

In the assignment

```
> X <- cbind(arg_1, arg_2, arg_3, ...)
```

the arguments to cbind() must be either vectors of any length, or matrices with the same column size, that is the same number of rows. The result is a matrix with the concatenated arguments *arg_1*, *arg_2*, ... forming the columns.

If some of the arguments to cbind() are vectors they may be shorter than the column size of any matrices present, in which case they are cyclically extended to match the matrix column size (or the length of the longest vector if no matrices are given).

The function rbind() does the corresponding operation for rows. In this case any vector arguments (possibly cyclically extended) are of course taken as row vectors.

Suppose X1 and X2 have the same number of rows. To combine these by columns into a matrix X, together with an initial column of ones we can use

```
> X <- cbind(1, X1, X2)
```

The result of rbind() or cbind() always has matrix status. Hence cbind(x) and rbind(x) are the simplest ways explicitly to allow the vector x to be treated as a column or row matrix respectively.

5.9 The concatenation function, c(), with arrays

It should be noted that whereas cbind() and rbind() are concatenation functions that respect dim attributes, the basic c() function does not—it clears numeric objects of all dim and dimnames attributes. This is occasionally useful in its own right.

The official way to coerce an array back to a simple vector object is to use as.vector()

```
> vec <- as.vector(X)
```

However, a similar result can be achieved by using c() with just one argument, simply for this side-effect:

```
> vec <- c(X)
```

There are slight differences between the two, but ultimately the choice between them is largely a matter of style (with the former being preferable).

5.10 Frequency tables from factors

Recall that a factor defines a partition into groups. Similarly, a pair of factors defines a two-way cross classification, and so on. The function table() allows frequency tables to be calculated from equal length factors. If there are k category arguments, the result is a k-way array of frequencies.

Suppose, for example, that statef is a factor giving the state code for each entry in a data vector. The assignment

```
> statefr <- table(statef)
```

gives in statefr a table of frequencies of each state in the sample. The frequencies are ordered and labelled by the levels attribute of the category. This simple case is equivalent to, but more convenient than,

```
> statefr <- tapply(statef, statef, length)
```

Further suppose that incomef is a category giving a suitably defined "income class" for each entry in the data vector, for example with the cut() function:

```
> factor(cut(incomes, breaks = 35+10*(0:7))) -> incomef
```

We can use the function table() to calculate a two-way table of frequencies:

```
> table(incomef,statef)
         statef
incomef    act nsw nt qld sa tas vic wa
  (35,45]   1   1  0   1  0   0   1  0
  (45,55]   1   1  1   1  2   0   1  3
  (55,65]   0   3  1   3  2   2   2  1
  (65,75]   0   1  0   0  0   0   1  0
```

Extension to higher-way frequency tables is immediate.

6 Lists and data frames

6.1 Lists

An R *list* is an object consisting of an ordered collection of objects, known as its *components*.

There is no particular need for the components to be of the same mode or type. For example, a list could consist of a numeric vector, a logical value, a matrix, a complex vector, a character array, a function, and so on. Here is a simple example of how to make a list:

```
> Lst <- list(name="Fred", wife="Mary", no.children=3,
              child.ages=c(4,7,9))
```

Components are always *numbered* and may always be referred to as such. Thus if `Lst` is the name of a list with four components, these may be individually referred to as `Lst[[1]]`, `Lst[[2]]`, `Lst[[3]]` and `Lst[[4]]`. If, further, `Lst[[4]]` is a vector subscripted array then `Lst[[4]][1]` is its first entry.

If `Lst` is a list, then the function `length(Lst)` gives the number of (top-level) components it has.

Components of lists may also be *named*, and in this case the component may be referred to either by giving the component name as a character string in place of the number in double square brackets, or, more conveniently, by giving an expression of the form

```
> name$component_name
```

for the same thing.

This is a very useful convention as it makes it easier to get the right component if you forget the number.

So in the simple example given above:

`Lst$name` is the same as `Lst[[1]]` and is the string `"Fred"`,

`Lst$wife` is the same as `Lst[[2]]` and is the string `"Mary"`,

`Lst$child.ages[1]` is the same as `Lst[[4]][1]` and is the number 4.

Additionally, one can also use the names of the list components in double square brackets, i.e., `Lst[["name"]]` is the same as `Lst$name`. This is especially useful, when the name of the component to be extracted is stored in another variable as in

```
> x <- "name"; Lst[[x]]
```

It is very important to distinguish `Lst[[1]]` from `Lst[1]`. '`[[...]]`' is the operator used to select a single element, whereas '`[...]`' is a general subscripting operator. Thus the former is the *first object in the list* `Lst`, and if it is a named list the name is *not* included. The latter is a *sublist of the list* `Lst` *consisting of the first entry only. If it is a named list, the name is transferred to the sublist.*

The names of components may be abbreviated down to the minimum number of letters needed to identify them uniquely. Thus the component `Lst$coefficients` may be minimally specified as `Lst$coe` and `Lst$covariance` as `Lst$cov`.

The vector of names is in fact simply an attribute of the list like any other and may be handled as such. Other structures besides lists may, of course, similarly be given a *names* attribute also.

6.2 Constructing and modifying lists

New lists may be formed from existing objects by the function `list()`. An assignment of the form

```
> Lst <- list(name_1=object_1, ..., name_m=object_m)
```

sets up a list `Lst` of m components using *object_1*, ..., *object_m* for the components and giving them names as specified by the argument names (which can be freely chosen). If these names are omitted, the components are numbered only. The components used to form the list are *copied* when forming the new list and the originals are not affected.

Lists, like any subscripted object, can be extended by specifying additional components. For example,

```
> Lst[5] <- list(matrix=Mat)
```

6.2.1 Concatenating lists

When the concatenation function `c()` is given list arguments, the result is an object of mode list also, whose components are those of the argument lists joined together in sequence.

```
> list.ABC <- c(list.A, list.B, list.C)
```

Recall that with vector objects as arguments the concatenation function similarly joined together all arguments into a single vector structure. In this case all other attributes, such as `dim` attributes, are discarded.

6.3 Data frames

A *data frame* is a list with class `"data.frame"`. There are restrictions on lists that may be made into data frames, namely

- The components must be vectors (numeric, character, or logical), factors, numeric matrices, lists, or other data frames.
- Matrices, lists, and data frames provide as many variables to the new data frame as they have columns, elements, or variables, respectively.
- Numeric vectors, logicals and factors are included as is, and character vectors are coerced to be factors, whose levels are the unique values appearing in the vector.
- Vector structures appearing as variables of the data frame must all have the *same length*, and matrix structures must all have the same *row size*.

A data frame may for many purposes be regarded as a matrix with columns possibly of differing modes and attributes. It may be displayed in matrix form, and its rows and columns extracted using matrix indexing conventions.

6.3.1 Making data frames

Objects satisfying the restrictions placed on the columns (components) of a data frame may be used to form one using the function `data.frame`:

```
> accountants <- data.frame(home=statef,
                            loot=incomes,
                            shot=incomef)
```

A list whose components conform to the restrictions of a data frame may be *coerced* into a data frame using the function `as.data.frame()`

The simplest way to construct a data frame from scratch is to use the `read.table()` function to read an entire data frame from an external file. This is discussed further in Chapter 7 [Reading data from files], page 47.

6.3.2 `attach()` and `detach()`

The $ notation, such as `accountants$statef`, for list components is not always very convenient. A useful facility would be somehow to make the components of a list or data frame temporarily visible as variables under their component name, without the need to quote the list name explicitly each time.

The `attach()` function, as well as having a directory name as its argument, may also have a data frame. Thus suppose `lentils` is a data frame with three variables `lentils$u`, `lentils$v`, `lentils$w`. The attach command

```
> attach(lentils)
```

places the data frame in the search path at position 2, and provided there are no variables `u`, `v` or `w` in position 1, `u`, `v` and `w` are available as variables from the data frame in their own right. At this point, an assignment such as

```
> u <- v+w
```

does not replace the component u of the data frame, but rather masks it with another variable u in the working directory at position 1 on the search path. To make a permanent change to the data frame itself, the simplest way is to resort once again to the $ notation:

```
> lentils$u <- v+w
```

However, the new value of component u is not visible until the data frame is detached and attached again.

To detach a data frame, use the function

```
> detach()
```

More precisely, this statement detaches from the search path the entity currently at position 2. Thus in the present context the variables u, v and w would be no longer visible, except under the list notation as `lentils$u` and so on. Entities at positions greater than 2 on the search path can be detached by giving their number to `detach`, but it is much safer to always use a name, for example by `detach(lentils)` or `detach("lentils")`

6.3.3 Working with data frames

A useful convention that allows you to work with many different problems comfortably together in the same working directory is

- gather together all variables for any well defined and separate problem in a data frame under a suitably informative name;
- when working with a problem attach the appropriate data frame at position 2, and use the working directory at level 1 for operational quantities and temporary variables;
- before leaving a problem, add any variables you wish to keep for future reference to the data frame using the $ form of assignment, and then `detach()`;
- finally remove all unwanted variables from the working directory and keep it as clean of left-over temporary variables as possible.

In this way it is quite simple to work with many problems in the same directory, all of which have variables named x, y and z, for example.

6.3.4 Attaching arbitrary lists

`attach()` is a generic function that allows not only directories and data frames to be attached to the search path, but other classes of object as well. In particular, any object of mode `"list"` may be attached in the same way:

```
> attach(any.old.list)
```

Anything that has been attached can be detached by detach, by position number or, preferably, by name.

6.3.5 Managing the search path

The function search shows the current search path and so is a very useful way to keep track of which data frames, lists and packages have been attached and detached. Initially, it shows the default path

```
> search()
".GlobalEnv"       "package:methods"    "package:stats"
"package:utils"    "package:graphics"   "Autoloads"
"package:base"
```

where .GlobalEnv is the workspace.[1]

After lentils is attached we have

```
> search()
".GlobalEnv"       "lentils"          "package:methods"
"package:stats"    "package:utils"    "package:graphics"
"Autoloads"        "package:base"
> ls(2)
"u" "v" "w"
```

and as we see ls (or objects) can be used to examine the contents of any position on the search path.

Finally, we detach the data frame and confirm it has been removed from the search path.

```
> detach("lentils")
> search()
".GlobalEnv"       "package:methods"    "package:stats"
"package:utils"    "package:graphics"   "Autoloads"
"package:base"
```

---

(1) See the R Reference Manual or the help for autoload for details of the "Autoloads" entry in the search path.

7 Reading data from files

Large data objects will usually be read as values from external files rather than entered during an R session at the keyboard. R input facilities are simple and their requirements are fairly strict and even rather inflexible. There is a clear presumption by the designers of R that you will be able to modify your input files using other tools, such as file editors or Perl[1], to fit in with the requirements of R. Generally this is very simple.

If variables are to be held mainly in data frames, as we strongly suggest they should be, an entire data frame can be read directly with the `read.table()` function. There is also a more primitive input function, `scan()`, that can be called directly.

For more details on importing data into R and also exporting data, see the *R Data Import/Export* manual.

7.1 The `read.table()` function

To read an entire data frame directly, the external file will normally have a special form:

- The first line of the file should have a *name* for each variable in the data frame.
- Each additional line of the file has a *row label* as its first item, followed by the values for each variable.

If the file has one fewer item in its first line than in its second, this arrangement is presumed to be in force. So the first few lines of a file to be read as a data frame might look as follows:

Input file form with names and row labels:

```
     Price    Floor     Area   Rooms    Age   Cent.heat
01   52.00    111.0      830     5      6.2      no
02   54.75    128.0      710     5      7.5      no
03   57.50    101.0     1000     5      4.2      no
04   57.50    131.0      690     6      8.8      no
05   59.75     93.0      900     5      1.9      yes
...
```

[1] Under UNIX, the utilities Sed or Awk can also be used.

By default, numeric items (except row labels) are read as numeric variables and non-numeric variables, such as `Cent.heat` in the example above, as factors. This can be changed if necessary.

The function `read.table()` can then be used to read the data frame directly:

```
> HousePrice <- read.table("houses.data")
```

Often you will want to avoid using the row labels from the file and use the default labels instead. In this case, the file may omit the row label column, as in the following:

Input file form without row labels:

```
Price    Floor     Area   Rooms     Age  Cent.heat
52.00    111.0      830     5       6.2       no
54.75    128.0      710     5       7.5       no
57.50    101.0     1000     5       4.2       no
57.50    131.0      690     6       8.8       no
59.75     93.0      900     5       1.9      yes
...
```

The data frame may then be read as

```
> HousePrice <- read.table("houses.data", header=TRUE)
```

where the `header=TRUE` option specifies that the first line is a line of headings, and hence, by implication from the form of the file, that no explicit row labels are given.

7.2 The `scan()` function

Suppose the data vectors are of equal length and are to be read in parallel. Further suppose that there are three vectors, the first of mode character and the remaining two of mode numeric, and the file is '`input.dat`'. The first step is to use `scan()` to read in the three vectors as a list, as follows

```
> inp <- scan("input.dat", list("",0,0))
```

The second argument is a dummy list structure that establishes the mode of the three vectors to be read. The result, held in `inp`, is a list whose components are the three vectors read in. To separate the data items into three separate vectors, use assignments like

```
> label <- inp[[1]]; x <- inp[[2]]; y <- inp[[3]]
```

More conveniently, the dummy list can have named components, in which case the names can be used to access the vectors read in. For example,

```
> inp <- scan("input.dat", list(id="", x=0, y=0))
```

If you wish to access the variables separately they may either be re-assigned to variables in the working frame:

```
> label <- inp$id; x <- inp$x; y <- inp$y
```

or the list may be attached at position 2 of the search path (see Section 6.3.4 [Attaching arbitrary lists], page 44).

If the second argument is a single value and not a list, a single vector is read in, all components of which must be of the same mode as the dummy value:

```
> X <- matrix(scan("light.dat", 0), ncol=5, byrow=TRUE)
```

There are other more elaborate input functions available, and these are detailed in the R Reference Manual.

7.3 Accessing builtin datasets

Over fifty datasets are supplied with R, and others are available in packages (including the standard packages supplied with R). Unlike S-PLUS, these datasets have to be loaded explicitly, using the function `data`. To see the list of datasets in the base system use

```
> data()
```

and to load one of these use, for example,

```
> data(infert)
```

In most cases, this will load an R object of the same name (usually a data frame). However, in a few cases it loads several objects, so check the R Reference Manual or help for the object to know what to expect.

7.3.1 Loading data from other R packages

To access data from another package, use the `package` argument. For example,

```
> data(package="stats")
> data(Puromycin, package="stats")
```

If a package has been attached by `library`, its datasets are automatically included in the search, so that

```
> library(stats) # normally attached
> data()
> data(Puromycin)
```

will list all the datasets in all the currently attached packages (at least **base** and **stats**) and then load `Puromycin` from the first package in which such a dataset is found.

User-contributed packages can be a rich source of datasets.

7.4 Editing data

When invoked on a data frame or matrix, `edit` brings up a separate spreadsheet-like environment for editing. This is useful for making small changes once a data set has been read. The command

```
> xnew <- edit(xold)
```

will allow you to edit your data set `xold`, and on completion the result is assigned to `xnew`. Note that the original object `xold` itself is not modified.

Use

```
> xnew <- edit(data.frame())
```

to enter new data via the spreadsheet interface.

8 Probability distributions

8.1 R as a set of statistical tables

One convenient use of R is to provide a comprehensive set of statistical tables. Functions are provided to evaluate the probability density function, the cumulative distribution function $P(X \leq x)$ and the quantile function (given q, the smallest x such that $P(X \leq x) > q$), and to simulate from a distribution. The following distributions are supported:

Distribution	R name	additional arguments
beta	`beta`	`shape1, shape2, ncp`
binomial	`binom`	`size, prob`
Cauchy	`cauchy`	`location, scale`
chi-squared	`chisq`	`df, ncp`
exponential	`exp`	`rate`
F	`f`	`df1, df1, ncp`
gamma	`gamma`	`shape, scale`
geometric	`geom`	`prob`
hypergeometric	`hyper`	`m, n, k`
log-normal	`lnorm`	`meanlog, sdlog`
logistic	`logis`	`location, scale`
negative binomial	`nbinom`	`size, prob`
normal	`norm`	`mean, sd`
Poisson	`pois`	`lambda`
Student's t	`t`	`df, ncp`
uniform	`unif`	`min, max`
Weibull	`weibull`	`shape, scale`
Wilcoxon	`wilcox`	`m, n`

To use these distributions, prefix the name given above by 'd' for the density function, 'p' for the CDF, 'q' for the quantile function and 'r' for the simulation function (*r*andom deviates).

For example, the function `dchisq()` computes the density of the chi-squared distribution at a given point x,

```
> dchisq(3.0, df = 4)
0.1673476
```

The first argument gives the value of x, and the named argument `df` specifies the number of degrees of freedom for the chi-squared distribution.

Similarly, the function pt() computes the cumulative distribution function (p-value) for the t-distribution,

```
> pt(-2.43, df = 13)
0.01516545
```

The first argument of the CDF functions gives the cut-off for the tail. By default, the cumulative distribution function is based on the lower tail of the function.

To calculate a p-value for the upper tail, specify the option lower.tail=FALSE:

```
> pt(-2.43, df = 13, lower.tail=FALSE)
0.9848346
```

Note that this value and the previous result correctly sum to 1.0, allowing for the numerical rounding of the displayed output.

For symmetric distributions, you can compute a 2-tailed p-value by simple multiplication,

```
> 2*pt(-2.43, df = 13)
0.0303309
```

In other cases, the 2-tailed p-value can be found by combining the values of the appropriate upper and lower tails.

The q-prefix functions compute quantile values. For example, the following expression computes the upper 1% point for an F(2, 7) distribution

```
> qf(0.99, 2, 7)
9.546578
```

The quantile functions are the inverse of the corresponding cumulative distribution functions, as can be seen by comparing the following result with the value of pt() above

```
> qt(0.01516545, df = 13)
-2.43
```

The first argument specifies the cumulative probability to be inverted. The quantile functions default to calculations using the lower tail, but respect the setting lower.tail=FALSE for calculations using the upper tail (in the same way as the cumulative distribution functions).

Finally, the r-prefix simulation functions are used for generating random variates. The following command will generate three random variates from the normal distribution $N(100, 5)$,

```
> rnorm(3, mean = 100, sd = 5)
92.69948 101.09035 99.39913
```

The first argument specifies the number of samples to be generated, and can also be set as the named argument 'n='. For the functions rhyper and rwilcox the name nn is used—to avoid confusion with other parameters of those distributions.

To generate uniform random variates in the range $[0, 1]$, call the function `runif`,

```
> runif(4)
0.9629671 0.4886081 0.1829479 0.8181007
```

The sequences of random variates produced by these functions can be controlled with a seed value (an integer), using the function `set.seed(seed)`:

```
> set.seed(12345)
> runif(4)
0.7209039 0.8757732 0.7609823 0.8861246
> runif(1)
0.4564810
> set.seed(12345)  # reset seed
> runif(4)
0.7209039 0.8757732 0.7609823 0.8861246
```

Initially, there is no seed; a new one is created from the current time (as an integer) when one is required. Hence, different sessions will give different simulation results, by default, unless the same seed is given.

The cumulative distribution and quantile functions, `pname` and `qname`, all have logical arguments `lower.tail` and `log.p` and the `dname` ones have `log`. This provides a way to obtain a reliable value of the cumulative (or "integrated") *hazard* function, $H(t) = -\log(1 - F(t))$, without loss of precision by

```
- pname(t, ..., lower.tail = FALSE, log.p = TRUE)
```

or more accurate log-likelihoods (by `dname(..., log = TRUE)`), directly.

In addition, there are functions `ptukey` and `qtukey` for the distribution of the studentized range of samples from a normal distribution.

The non-centrality parameter `ncp` is only applicable to some functions: see the R Reference Manual or help pages for details.

8.2 Examining the distribution of a set of data

Given a (univariate) set of data, we can examine its distribution in a large number of ways. The simplest is to examine the numbers. Two slightly different summaries are given by `summary` and `fivenum`, which returns Tukey's five number summary (minimum, lower-hinge, median, upper-hinge, maximum) for the input data:

```
> data(faithful)
> attach(faithful)
> summary(eruptions)
   Min. 1st Qu.  Median    Mean 3rd Qu.    Max.
  1.600   2.163   4.000   3.488   4.454   5.100
```

```
> fivenum(eruptions)
1.6000 2.1585 4.0000 4.4585 5.1000
```

A display of the numbers is produced by `stem` (a "stem and leaf" plot):

```
> stem(eruptions)

  The decimal point is 1 digit(s) to the left of the |

  16 | 070355555588
  18 | 000022233333335577777777888822335777888
  20 | 00002223378800035778
  22 | 0002335578023578
  24 | 00228
  26 | 23
  28 | 080
  30 | 7
  32 | 2337
  34 | 250077
  36 | 0000823577
  38 | 2333335582225577
  40 | 0000003357788888002233555577778
  42 | 03335555778800233333555577778
  44 | 02222335557780000000023333357778888
  46 | 0000233357700000023578
  48 | 00000022335800333
  50 | 0370
```

A stem-and-leaf plot is like a histogram, and R has a function `hist` to plot histograms.

```
> hist(eruptions)
## make the bins smaller, make a plot of density
> hist(eruptions, seq(1.6, 5.2, 0.2), prob=TRUE)
> lines(density(eruptions, bw=0.1))
> rug(eruptions) # show the actual data points
```

More elegant density plots can be made by `density`, and we added a line produced by `density` in this example. The bandwidth `bw` was chosen by trial-and-error as the default gives too much smoothing (it usually does for "interesting" densities). Automated methods of bandwidth choice are implemented in the packages **MASS** and **KernSmooth**.

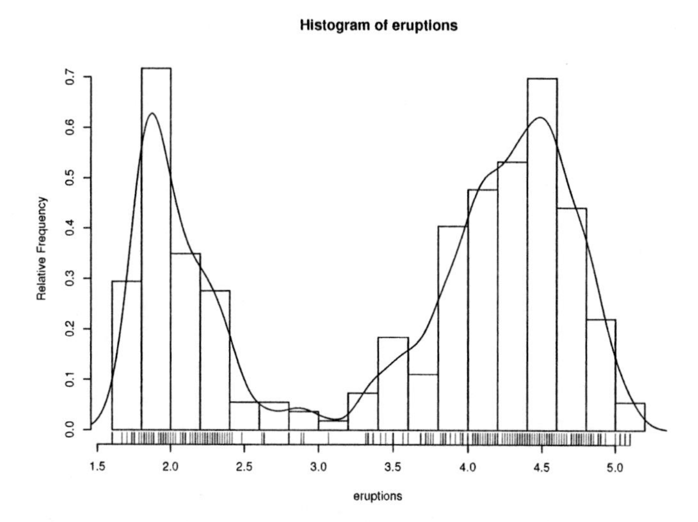

We can plot the empirical cumulative distribution function using the function `ecdf` in the standard package **stats**.(1)

```
> library(stepfun)  # for older versions of R
> plot(ecdf(eruptions), do.points=FALSE, verticals=TRUE)
```

This distribution is obviously far from any standard distribution. How about the right-hand mode, say eruptions of longer than 3 minutes? Let us fit a normal distribution and overlay the fitted CDF.

```
> long <- eruptions[eruptions > 3]
> plot(ecdf(long), do.points=FALSE, verticals=TRUE)
> x <- seq(3, 5.4, 0.01)
> lines(x,
        pnorm(x, mean=mean(long), sd=sqrt(var(long))),
        lty=3)
```

(1) In older versions of R, the function `ecdf` is in the package **stepfun** and can be loaded with the command `library(stepfun)`.

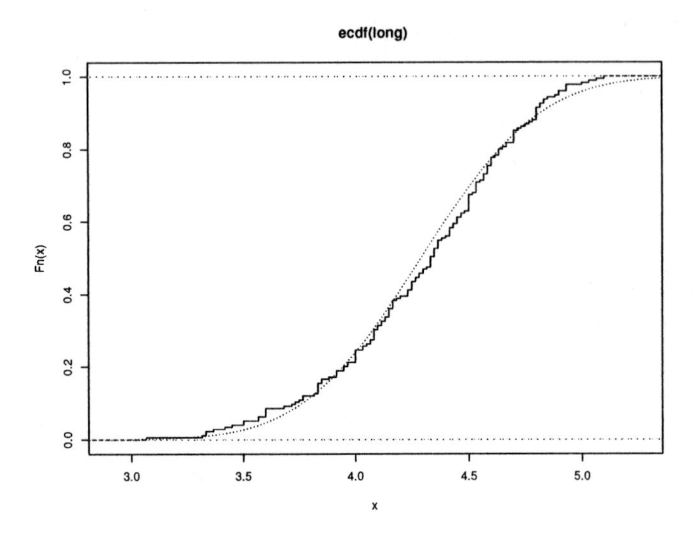

Quantile-quantile (Q-Q) plots can help us examine this more carefully:

```
> par(pty="s")
> qqnorm(long); qqline(long)
```

This shows a reasonable fit but a shorter right tail than one would expect from a normal distribution:

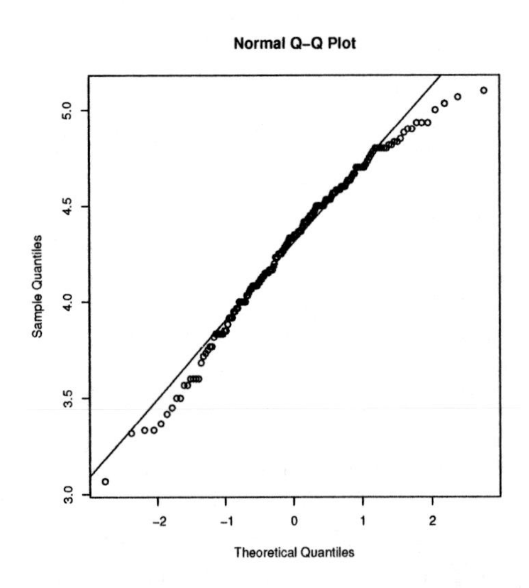

Let us compare this with some simulated data from a t distribution,

```
> x <- rt(250, df = 5)
> qqnorm(x); qqline(x)
```

This will usually (if it is a random sample) show longer tails than expected for a normal. We can make a Q-Q plot against the generating distribution by

```
> qqplot(qt(ppoints(250), df=5), x,
         xlab="Q-Q plot for t dsn")
> qqline(x)
```

Finally, we might want a more formal test of agreement with normality (or not). The standard package **stats** provides the Shapiro-Wilk test,

```
> shapiro.test(long)
         Shapiro-Wilk normality test
data:  long
W = 0.9793, p-value = 0.01052
```

and the Kolmogorov-Smirnov test,

```
> ks.test(long, "pnorm", mean=mean(long),
          sd=sqrt(var(long)))
         One-sample Kolmogorov-Smirnov test
data:  long
D = 0.0661, p-value = 0.4284
alternative hypothesis: two.sided
```

(Note that the distribution theory is not valid here as we have estimated the parameters of the normal distribution from the same sample.)

8.3 One- and two-sample tests

So far we have compared a single sample to a normal distribution. A much more common operation is to compare aspects of two samples. Note that in R, all "classical" tests including the ones used below are in the package **stats**.

Consider the following sets of data on the latent heat of the fusion of ice (*cal/gm*) from Rice (1995, p.490)

```
Method A: 79.98 80.04 80.02 80.04 80.03
          80.03 80.04 79.97 80.05 80.03
          80.02 80.00 80.02
Method B: 80.02 79.94 79.98 79.97 79.97
          80.03 79.95 79.97
```

The data can be entered at the keyboard using `scan()`, which reads input one line at time. To finish entering data, hit ⟨ENTER⟩ twice.

```
> A <- scan()
79.98 80.04 80.02 80.04 80.03 80.03 80.04 79.97
```

```
80.05 80.03 80.02 80.00 80.02 ⟨ENTER⟩
⟨ENTER⟩

> B <- scan()
80.02 79.94 79.98 79.97 79.97 80.03 79.95 79.97 ⟨ENTER⟩
⟨ENTER⟩
```

Boxplots provide a simple graphical comparison of the two samples,

```
> boxplot(A, B)
```

which indicates that the first group tends to give higher results than the second:

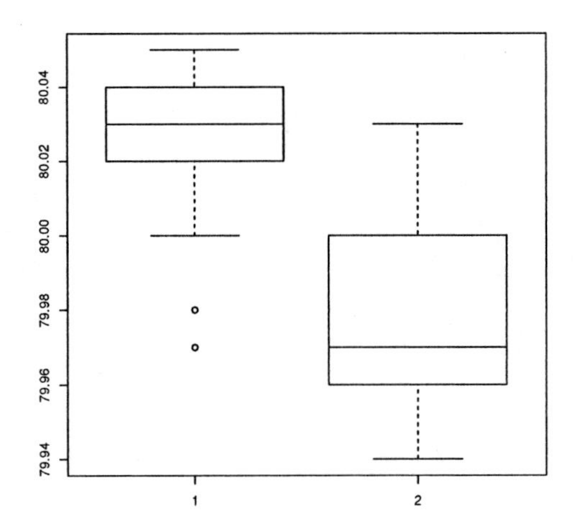

To test for the equality of the means of the two examples, we can use an *unpaired t*-test by

```
> t.test(A, B)
     Welch Two Sample t-test
data:  A and B
t = 3.2499, df = 12.027, p-value = 0.00694
alternative hypothesis: true difference in
  means is not equal to 0
95 percent confidence interval:
 0.01385526 0.07018320
sample estimates:
mean of x mean of y
 80.02077  79.97875
```

which does indicate a significant difference, assuming normality. By default, the R function does not assume equality of variances in the two samples (in contrast to the similar S-PLUS `t.test` function). We can use the F test to test for equality in the variances, provided that the two samples are from normal populations:

```
> var.test(A, B)
     F test to compare two variances
data:  A and B
F = 0.5837, num df = 12, denom df = 7, p-value = 0.3938
alternative hypothesis: true ratio of variances
  is not equal to 1
95 percent confidence interval:
 0.1251097 2.1052687
sample estimates:
ratio of variances
         0.5837405
```

which shows no evidence of a significant difference, and so we can use the classical t-test that assumes equality of the variances:

```
> t.test(A, B, var.equal=TRUE)
     Two Sample t-test
data:  A and B
t = 3.4722, df = 19, p-value = 0.002551
alternative hypothesis: true difference in means
  is not equal to 0
95 percent confidence interval:
 0.01669058 0.06734788
sample estimates:
mean of x mean of y
 80.02077  79.97875
```

All these tests assume normality of the two samples. The two-sample Wilcoxon (or Mann-Whitney) test only assumes a common continuous distribution under the null hypothesis:

```
> wilcox.test(A, B)
     Wilcoxon rank sum test with continuity correction
data:  A and B
W = 89, p-value = 0.007497
alternative hypothesis: true mu is not equal to 0
Warning message:
Cannot compute exact p-value with
  ties in: wilcox.test(A, B)
```

Note the warning: there are several ties in each sample, which suggests strongly that these data are from a discrete distribution (probably due to rounding).

There are several ways to compare graphically the two samples. We have already seen a pair of boxplots. The following

```
> library(stepfun)  # for older versions of R
> plot(ecdf(A), do.points=FALSE, verticals=TRUE,
       xlim=range(A, B))
> plot(ecdf(B), do.points=FALSE, verticals=TRUE,
       add=TRUE)
```

will show the two empirical CDFs, and `qqplot` will perform a Q-Q plot of the two samples. The Kolmogorov-Smirnov test is of the maximal vertical distance between the two ecdf's, assuming a common continuous distribution:

```
> ks.test(A, B)
        Two-sample Kolmogorov-Smirnov test
data:  A and B
D = 0.5962, p-value = 0.05919
alternative hypothesis: two.sided
Warning message:
cannot compute correct p-values with
  ties in: ks.test(A, B)
```

9 Grouping, loops and conditional execution

9.1 Grouped expressions

R is an expression language, in the sense that its only command type is a function or expression which returns a result. Even an assignment is an expression whose result is the value assigned, and it may be used wherever any expression may be used; in particular, multiple assignments such as `x <- y <- z <- 0` are possible.

Commands may be grouped together in braces, `{`*expr_1* `;` ... `;` *expr_m*`}`, in which case the value of the group is the result of the last expression in the group evaluated. Since such a group is also an expression it may, for example, be itself included in parentheses and used a part of an even larger expression, and so on.

9.2 Control statements

9.2.1 Conditional execution: `if` statements

The R language provides the standard conditional construction '`if`', with the form

```
> if (expr_1) expr_2 else expr_3
```

where *expr_1* must evaluate to a logical value, and the result of the entire expression is then either *expr_2* or *expr_3* respectively.

The "short-circuit" operators `&&` and `||` are often used as the condition in an `if` statement. Whereas `&` and `|` apply element-wise to vectors, `&&` and `||` apply to vectors of length one, and only evaluate their second argument if necessary.

There is a vectorized version of the `if/else` construct, the `ifelse` function. This has the form `ifelse(condition, a, b)` and returns a vector of the length of its longest argument, with elements `a[i]` if `condition[i]` is true, otherwise `b[i]`.

9.2.2 Repetitive execution: `for` loops, `repeat` and `while`

There is also a `for` loop construction which has the form

```
> for (name in expr_1) expr_2
```

where *name* is the loop variable, *expr_1* is a vector expression (often a sequence like 1:20), and *expr_2* is often a grouped expression with its sub-expressions written in terms of the dummy *name*. The body of the loop, *expr_2*, is repeatedly evaluated as *name* ranges through the values in the vector result of *expr_1*.

As an example, suppose ind is a vector of class indicators and we wish to produce separate plots of y versus x within classes. One possibility here is to use coplot() (to be discussed later) which will produce an array of plots corresponding to each level of the factor. Another way to do this, now putting all plots on the one display, is as follows:

```
> xc <- split(x, ind)
> yc <- split(y, ind)
> for (i in 1:length(yc)) {
    plot(xc[[i]], yc[[i]]);
    abline(lsfit(xc[[i]], yc[[i]]))
  }
```

The function split() splits its first argument into a list of vectors, according to the classes specified by a category. This is a useful function, mostly used in connection with boxplots. See the R Reference Manual or the help pages for further details.

> **WARNING**: for() loops are used in R code much less often than in compiled languages. Code that takes a 'whole object' view is likely to be both clearer and faster in R.

Other looping constructs include the 'repeat' statement,

```
> repeat expr
```

and the 'while' loop,

```
> while (condition) expr
```

The break statement can be used to terminate any loop, possibly abnormally. This is the only way to terminate repeat loops.

The next statement can be used to discontinue one particular cycle and skip to the "next".

Control statements are most often used in connection with *functions* which are discussed in Chapter 10 [Writing your own functions], page 63, where more examples will emerge.

10 Writing your own functions

As we have seen informally along the way, the R language allows the user to create objects of mode *function*. These are true R functions that are stored in a special internal form and may be used in further expressions and so on. In the process, the language gains enormously in power, convenience and elegance, and learning to write useful functions is one of the main ways to make your use of R comfortable and productive.

It should be emphasized that most of the functions supplied as part of the R system, such as `mean()`, `var()`, `postscript()` and so on, are themselves written in R and thus do not differ materially from user written functions.

A function is defined by an assignment of the form

```
> name <- function(arg_1, arg_2, ...) expression
```

The *expression* is an R expression (usually a grouped expression) that uses the arguments, *arg_i*, to calculate a value. The value of this expression is the value returned by the function. A call to a function takes the form `name(expr_1, expr_2, ...)` and may occur anywhere an expression is legitimate.

10.1 Simple examples

As a first example, consider a function to calculate the two sample *t*-statistic, showing "all the steps". This is an artificial example, of course, since there are other, simpler ways of achieving the same end.

The function is defined as follows:

```
> twosam <- function(y1, y2) {
    n1  <- length(y1); n2  <- length(y2)
    yb1 <- mean(y1);   yb2 <- mean(y2)
    s1  <- var(y1);    s2  <- var(y2)
    s <- ((n1-1)*s1 + (n2-1)*s2)/(n1+n2-2)
    tst <- (yb1 - yb2)/sqrt(s*(1/n1 + 1/n2))
    tst
  }
```

With this function defined, you could perform two sample *t*-tests using a call such as

```
> tstat <- twosam(data$male, data$female); tstat
```

The definition of an existing function can be displayed by entering the function name without parentheses or arguments:

```
> twosam
function(y1, y2) {
    n1  <- length(y1); n2  <- length(y2)
    yb1 <- mean(y1);   yb2 <- mean(y2)
    s1  <- var(y1);    s2  <- var(y2)
    s <- ((n1-1)*s1 + (n2-1)*s2)/(n1+n2-2)
    tst <- (yb1 - yb2)/sqrt(s*(1/n1 + 1/n2))
    tst
}
```

Note that displaying the definition of a function does not execute it. For example, omitting the parentheses () from a command such as help() shows its definition, instead of executing the actual help() command. The internal working of many standard R commands can be examined in this way, and is useful for checking the details of their algorithms.

As a second example, consider a function to emulate directly the MATLAB backslash command, which returns the coefficients of the orthogonal projection of the vector y onto the column space of the matrix, X. This is ordinarily called the least squares estimates of the regression coefficients, and would be computed with the qr() function; however this is sometimes a bit tricky to call qr() directly and it pays to have a simple function such as the following to use it safely.

Given an n-by-1 vector y and an n-by-p matrix X, the backslash operator $X \setminus y$ is defined as $(X'X)^{-}X'y$, where $(X'X)^{-}$ is a generalized inverse of $X'X$.

```
> bslash <- function(X, y) {
    X <- qr(X)
    qr.coef(X, y)
  }
```

This function definition is short enough to be entered directly at the R prompt. For longer functions, you can create or modify any object in R using the command fix(), which starts an interactive editor. For example,

```
> fix(bslash)
```

will start an editor on the function bslash, or allow you to enter a new definition if bslash does not yet exist.

After this object is created, it is permanent and may be used in statements such as

```
> regcoeff <- bslash(Xmat, yvar)
```

and so on.

The classical R function `lsfit()` does this job quite well, and more[1]. It uses the functions `qr()` and `qr.coef()` in the slightly counterintuitive way above to do this part of the calculation. Hence there is probably some value in having just this part isolated in a simple function if it is going to be in frequent use. If so, we may wish to make it a matrix binary operator for even more convenient use.

10.2 Defining new binary operators

Had we given the `bslash()` function a different name, namely one of the form

```
%anything%
```

it could have been used as a *binary operator* in expressions rather than in function form. Suppose, for example, we choose `!` for the internal character. The function definition would then start as

```
> "%!%" <- function(X, y) { ... }
```

The function could then be used as `X %!% y`. Note the use of quote marks around the operator. We use the exclamation mark ('`!`') as the operator in this case, since the backslash symbol itself is not a convenient choice—it is the escape character inside a string.

The matrix multiplication operator `%*%` and the outer product matrix operator `%o%` are other examples of binary operators defined in this way.

10.3 Named arguments and defaults

As first noted in Section 2.3 [Generating regular sequences], page 15, if the arguments to a called function are given in the "*name=object*" form, they may be given in any order. Furthermore, the argument sequence may begin in the unnamed, positional form, and specify named arguments after the positional arguments.

Thus, if there is a function `fun1` defined by

```
> fun1 <- function(data, data.frame, graph, limit) {
    [function body omitted]
  }
```

then the function may be invoked in several ways, for example

```
> ans <- fun1(d, df, TRUE, 20)
```

and,

```
> ans <- fun1(d, df, graph=TRUE, limit=20)
```

and,

---

[1] See also the methods described in Chapter 11 [Statistical models in R], page 75

```
> ans <- fun1(data=d, limit=20, graph=TRUE,
                    data.frame=df)
```

are all equivalent.

Arguments can be given default values, in which case they may be omitted altogether from the call when the defaults are appropriate. For example, if `fun1` were defined as

```
> fun1 <- function(data, data.frame, graph=TRUE,
            limit=20) { ... }
```

it could be called as

```
> ans <- fun1(d, df)
```

which is now equivalent to the three cases above, or as

```
> ans <- fun1(d, df, limit=10)
```

which changes one of the defaults.

It is important to note that defaults may be arbitrary expressions, even involving other arguments to the same function; they are not restricted to be constants as in our simple example here.

10.4 The '...' argument

Another frequent requirement is to allow one function to pass on argument settings to another. For example, many graphics functions use the function `par()` and functions like `plot()` allow the user to pass on graphical parameters to `par()` to control the graphical output. (See Section 12.4.1 [The par() function], page 101, for more details on the `par()` function). This can be done by including an extra argument, literally '`...`', of the function, which may then be passed on. An outline example is given below:

```
fun1 <- function(data, data.frame, graph=TRUE,
                        limit=20, ...) {
  [omitted statements]
  if (graph)
    par(pch="*", ...)
  [more omissions]
}
```

10.5 Assignments within functions

Note that *any ordinary assignments done within the function are local and temporary, and are lost after exit from the function.* Thus the assignment `X <- qr(X)` does not affect the value of the argument in the calling program.

To understand completely the rules governing the scope of R assignments the reader needs to be familiar with the notion of an evaluation *frame*. This is an advanced topic and is not covered further here. For more information, see the entries for `eval` and its related functions in the R Reference Manual or online help.

If global and permanent assignments are intended within a function, then either the "superassignment" operator `<<-` or the function `assign()` can be used—see the R Reference Manual or help pages for details. S-PLUS users should be aware that `<<-` has different semantics in R. These are discussed further in Section 10.7 [Scope], page 69.

10.6 More advanced examples

10.6.1 Efficiency factors in block designs

As a more complete, if a little pedestrian, example of a function, consider finding the efficiency factors for a block design. (Some aspects of this problem have already been discussed in Section 5.3 [Index arrays], page 30.)

A block design is defined by two factors, say `blocks` (`b` levels) and `varieties` (`v` levels). If R and K are the v by v and b by b *replications* and *block size* matrices, respectively, and N is the b by v incidence matrix, then the efficiency factors are defined as the eigenvalues of the matrix

$$E = I_v - R^{-1/2}N'K^{-1}NR^{-1/2} = I_v - A'A,$$

where $A = K^{-1/2}NR^{-1/2}$. One way to write the function is given below:

```
> bdeff <- function(blocks, varieties) {
    blocks <- as.factor(blocks)           # minor safety move
    b <- length(levels(blocks))
    varieties <- as.factor(varieties) # minor safety move
    v <- length(levels(varieties))
    K <- as.vector(table(blocks))         # remove dim attr
    R <- as.vector(table(varieties))  # remove dim attr
    N <- table(blocks, varieties)
    A <- 1/sqrt(K) * N * rep(1/sqrt(R), rep(b, v))
    sv <- svd(A)
    list(eff=1 - sv$d^2, blockcv=sv$u, varietycv=sv$v)
  }
```

The singular value decomposition is used instead of the eigenvalue routines, because it is numerically more stable in this case.

The result of the function is a list giving not only the efficiency factors as the first component, but also the block and variety canonical contrasts, since sometimes these give additional useful qualitative information.

10.6.2 Dropping all names in a printed array

When printing large matrices or arrays, it is often useful to display them in close block form, without the array names or numbers. Removing the `dimnames` attribute will not achieve this effect, but rather the array must be given a `dimnames` attribute consisting of empty strings. For example, to print a matrix, `X`

```
> temp <- X
> dimnames(temp) <- list(rep("", nrow(X)),
                         rep("", ncol(X)))
> temp; rm(temp)
```

This can be much more conveniently done using a function, `no.dimnames()` (shown below), as a "wrapper" to achieve the same result. It also illustrates how some effective and useful user functions can be quite short.

```
no.dimnames <- function(a) {
  ## Remove all dimension names from an array for compact printing.
  d <- list()
  j <- 0
  for(i in dim(a)) {
    d[[j <- j + 1]] <- rep("", i)
  }
  dimnames(a) <- d
  a
}
```

With this function defined, an array may be printed in close format using

```
> no.dimnames(X)
```

This is particularly useful for large integer arrays, where patterns are the real interest rather than the values.

10.6.3 Recursive numerical integration

Functions may be recursive, and can also define functions within themselves (nested functions). Note, however, that such functions, or indeed variables, are not inherited by called functions in higher evaluation frames as they would be if they were on the search path.

The example below shows a naive way of performing one-dimensional numerical integration using recursive function calls. The integrand is evaluated at the end points of the range and in the middle. If the one-panel trapezium rule answer is close enough to the two-panel, then the latter is returned as the value. Otherwise the same process is recursively applied to each panel.

The result is an adaptive integration process that concentrates function evaluations in regions where the integrand is farthest from linear. There is, however, a heavy overhead, and the function is only competitive with other algorithms when the integrand is both smooth and very difficult to evaluate. This example is also given partly as a little puzzle in R programming.

```
area <- function(f, a, b, eps = 1.0e-06, lim = 10) {
  fun1 <- function(f, a, b, fa, fb, a0, eps, lim, fun) {
    ## function 'fun1' is only visible inside 'area'
    d <- (a + b)/2
    h <- (b - a)/4
    fd <- f(d)
    a1 <- h * (fa + fd)
    a2 <- h * (fd + fb)
    if(abs(a0 - a1 - a2) < eps || lim == 0)
      return(a1 + a2)
    else {
      return(fun(f, a, d, fa, fd, a1, eps, lim - 1, fun) +
             fun(f, d, b, fd, fb, a2, eps, lim - 1, fun))
    }
  }
  fa <- f(a)
  fb <- f(b)
  a0 <- ((fa + fb) * (b - a))/2
  fun1(f, a, b, fa, fb, a0, eps, lim, fun1)
}
```

The following test cases demonstrate the use of this function:

```
> area(sqrt,0,1)
0.6666557
> testfn <- function(x) 1/(1+x^2)
> area(testfn, 0, 100)
1.560808
```

10.7 Scope

The discussion in this section is somewhat more technical than in other parts of this document. However, it details one of the major differences between S-PLUS and R.

The symbols which occur in the body of a function can be divided into three classes; formal parameters, local variables and free variables. The formal parameters of a function are those occurring in the argument list of the function. Their values are determined by the process of *binding* the actual function arguments to the formal parameters. Local variables are those whose values are determined by the evaluation of expressions in the body of the functions. Variables which are not formal parameters or local variables are called free variables. Free variables become local variables if they are assigned to. Consider the following function definition:

```
f <- function(x) {
  y <- 2*x
  print(x)
  print(y)
  print(z)
}
```

In this function, x is a formal parameter, y is a local variable and z is a free variable.

In R, the free variable bindings are resolved by first looking in the environment in which the function was created. This is called *lexical scope*. To demonstrate this, we define the following function called cube:

```
cube <- function(n) {
  sq <- function() n*n
  n*sq()
}
```

The variable n in the function sq is not an argument to that function. Therefore it is a free variable and the scoping rules must be used to ascertain the value that is to be associated with it.

Under static scope (S-PLUS) the value is that associated with a global variable named n. Under lexical scope (R) it is the parameter to the function cube since that is the active binding for the variable n at the time the function sq was defined.

The difference between evaluation in R and evaluation in S-PLUS is that S-PLUS looks for a global variable called n while R first looks for a variable called n in the environment created when cube was invoked.

```
## first evaluation in S
S> cube(2)
Error in sq(): Object "n" not found
Dumped
S> n <- 3
S> cube(2)
18
## then the same function evaluated in R
```

```
R> cube(2)
8
```

Lexical scope can also be used to give functions *mutable state*. In the following example we show how R can be used to mimic a bank account. A functioning bank account needs to have a balance or total, a function for making withdrawals, a function for making deposits and a function for stating the current balance. We achieve this by creating the three functions within `account` and then returning a list containing them. When `account` is invoked it takes a numerical argument `total` and returns a list containing the three functions. Because these functions are defined in an environment which contains `total`, they will have access to its value.

The special assignment operator `<<-` is used to change the value associated with `total`. This operator looks back in enclosing environments for an environment that contains the symbol `total` and when it finds such an environment it replaces the value, in that environment, with the value of right hand side. If the global or top-level environment is reached without finding the symbol `total` then that variable is created and assigned to there.

For most users `<<-` creates a global variable and assigns the value of the right hand side to it(2). Only when `<<-` has been used in a function that was returned as the value of another function will the special behavior described here occur.

```
open.account <- function(total) {
  list(
    deposit = function(amount) {
      if(amount <= 0)
        stop("Deposits must be positive!\n")
      total <<- total + amount
      cat(amount, "deposited. ",
          "Your balance is", total, "\n\n")
    },
    withdraw = function(amount) {
      if(amount > total)
        stop("You don't have that much money!\n")
      total <<- total - amount
      cat(amount, "withdrawn. ",
          "Your balance is", total, "\n\n")
    },
    balance = function() {
      cat("Your balance is", total, "\n\n")
```

(2) In some sense this mimics the behavior in S-PLUS since in S-PLUS this operator always creates or assigns to a global variable.

```
        }
      )
    }
```

The function can then be used as follows:

```
> ross <- open.account(100)
> robert <- open.account(200)

> ross$withdraw(30)
30 withdrawn.  Your balance is 70
> ross$balance()
Your balance is 70
> robert$balance()
Your balance is 200

> ross$deposit(50)
50 deposited.  Your balance is 120
> ross$balance()
Your balance is 120
> ross$withdraw(500)
Error in ross$withdraw(500) : You don't have that
  much money!
```

10.8 Customizing the environment

Users can customize their environment in several different ways. There is a site initialization file, and every directory can have its own individual initialization file. Finally, the special functions `.First` and `.Last` can be used.

The location of the site initialization file is taken from the value of the `R_PROFILE` environment variable. If that variable is unset, the file '`Rprofile.site`' in the R home subdirectory '`etc`' is used. This file should contain the commands that you want to execute every time R is started under your system. A second, personal, profile file named '`.Rprofile`'[3] can be placed in any directory. If R is invoked in that directory then that file will be sourced. This file gives individual users control over their workspace and allows for different startup procedures in different working directories. If no '`.Rprofile`' file is found in the startup directory, then R looks for a '`.Rprofile`' file in the user's home directory and uses that, if it exists.

Any function named `.First()` in either of the two profile files or in the '`.RData`' image has a special status. It is automatically executed at the

---

[3] Filenames beginning with a dot '`.`' are hidden by default under UNIX.

beginning of an R session and may be used to initialize the environment. For example, the definition in the example below alters the prompt to $ and sets up various other useful things that can then be taken for granted in the rest of the session.

Thus, the sequence in which files are executed is: '`Rprofile.site`', '`.Rprofile`', '`.RData`' and then `.First()`. A definition in later files will mask definitions in earlier files.

```
> .First <- function() {
# $ is the prompt
  options(prompt="$ ", continue="+\t")
# custom numbers and printout
  options(digits=5, length=999)
# my personal package
  source(file.path(Sys.getenv("HOME"), "R", "mystuff.R"))
}
```

Similarly, the function `.Last()` is executed at the very end of the session. An example is given below.

```
> .Last <- function() {
# a small safety measure.
  graphics.off()
# Is it time for lunch?
  cat(paste(date(),"\nAdios\n"))
}
```

10.9 Classes, generic functions and object orientation

The class of an object determines how it will be treated by what are known as *generic* functions. Put the other way round, a generic function performs a task or action on its arguments *specific to the class of the argument itself.* If the argument lacks any class attribute, or has a class not catered for specifically by the generic function in question, there is always a *default action* provided.

The class mechanism makes it possible to write generic functions for special purposes. An example makes things clearer: the generic function `plot()` displays different types of objects graphically according to their `class` attribute, `summary()` displays analyses of various types according to their class, and `anova()` compares different types of statistical models by their classes.

The number of generic functions that can treat a class in a specific way can be quite large. For example, the functions that can accommodate in some fashion objects of class `"data.frame"` include

```
[      [[<-    any     as.matrix
[<-    model   plot    summary
```

A complete list can be found using the `methods()` function:

```
> methods(class="data.frame")
```

Conversely, the number of classes a generic function can handle can also be quite large. For example, the `plot()` function has a default method and variants for objects of classes `"data.frame"`, `"density"`, `"factor"`, and more. A complete list can again be found using the `methods()` function:

```
> methods(plot)
```

The reader is referred to the official references for a complete discussion of this mechanism.

11 Statistical models in R

This section presumes the reader has some familiarity with statistical methodology, in particular with regression analysis and the analysis of variance. Later we will make some rather more ambitious presumptions, namely that something is known about generalized linear models and nonlinear regression.

The requirements for fitting statistical models are sufficiently well defined to make it possible to construct general tools that apply in a broad spectrum of problems.

R provides an interlocking suite of facilities that make fitting statistical models very simple. As we mentioned in the introduction, the basic output is minimal, and one needs to ask for details by calling extractor functions.

11.1 Defining statistical models; formulae

Models in R are described using a schematic notation, which gives the types of terms in the model in a short-hand form. This avoids the need to provide an explicit expression for calculating model values. The tilde operator ~ is used to define a *model formula* in R. The form, for an ordinary linear model, is

```
response ~ op_1 term_1 op_2 term_2 op_3 term_3 ...
```

where

response
: is a vector or matrix, (or expression evaluating to a vector or matrix) defining the response variable(s).

op_i
: is an operator, either + or -, implying the inclusion or exclusion of a term in the model, (the first is optional).

term_i
: is either

- a vector or matrix expression, or 1,
- a factor, or
- a *formula expression* consisting of factors, vectors or matrices connected by *formula operators*.

In all cases each term defines a collection of columns either to be added to or removed from the model matrix. A 1 stands for an intercept column and is by default included in the model matrix unless explicitly removed.

The *formula operators* are similar in effect to the Wilkinson and Rogers notation used by such programs as Glim and Genstat. One inevitable change is that the operator '.' becomes ':' since the period is a valid name character in R.

The notation is summarized below (based on Chambers & Hastie, 1992, p.29):

`Y ~ M`
: Y is modeled as M.

`M_1 + M_2`
: Include *M_1* and *M_2*.

`M_1 - M_2`
: Include *M_1* leaving out terms of *M_2*.

`M_1 : M_2`
: The tensor product of *M_1* and *M_2*. If both terms are factors, then the "subclasses" factor.

`M_1 %in% M_2`
: Similar to `M_1:M_2`, but with a different coding.

`M_1 * M_2`
: `M_1 + M_2 + M_1:M_2`.

`M_1 / M_2`
: `M_1 + M_2 %in% M_1`.

`M^n`
: All terms in M together with "interactions" up to order n

`I(M)`
: Insulate M. Inside M all operators have their normal arithmetic meaning, and that term appears in the model matrix.

Note that inside the parentheses that usually enclose function arguments all operators have their normal arithmetic meaning. The function `I()` is an identity function used only to allow terms in model formulae to be defined using arithmetic operators.

Note particularly that the model formulae specify the *columns of the model matrix*, the specification of the parameters being implicit. This is not the case in other contexts, for example in specifying nonlinear models.

The table above gives the formal specification of model formulae, a few examples may usefully set the picture.

Examples

Suppose `y`, `x`, `x0`, `x1`, `x2`, ... are numeric variables, `X` is a matrix and `A`, `B`, `C`, ... are factors. The following formulae below specify statistical models, as described on the right.

`y ~ x`
`y ~ 1 + x`
: Both these expressions imply the same simple linear regression model of y on x. The first has an implicit intercept term, and the second an explicit one.

`y ~ 0 + x`
`y ~ -1 + x`
`y ~ x - 1`
: Simple linear regression of y on x through the origin (that is, without an intercept term).

`log(y) ~ x1 + x2`
: Multiple regression of the transformed variable, $\log(y)$, on $x1$ and $x2$ (with an implicit intercept term).

`y ~ poly(x,2)`
`y ~ 1 + x + I(x^2)`
: Polynomial regression of y on x of degree 2. The first form uses orthogonal polynomials, and the second uses explicit powers, as basis.

`y ~ X + poly(x,2)`
: Multiple regression y with model matrix consisting of the matrix X as well as polynomial terms in x to degree 2.

`y ~ A`
: Single classification analysis of variance model of y, with classes determined by A.

`y ~ A + x`
: Single classification analysis of covariance model of y, with classes determined by A, and with covariate x.

`y ~ A*B`
`y ~ A + B + A:B`
`y ~ B %in% A`
`y ~ A/B`
: Two factor non-additive model of y on A and B. The first two specify the same crossed classification and the second two specify the same nested classification. In abstract terms, all four specify the same model subspace.

`y ~ (A + B + C)^2`
`y ~ A*B*C - A:B:C`
: Three factor experiment but with a model containing main effects and two factor interactions only. Both formulae specify the same model.

```
y ~ A * x
y ~ A/x
y ~ A/(1 + x) - 1
```

Separate simple linear regression models of y on x within the levels of A, with different codings. The last form produces explicit estimates of as many different intercepts and slopes as there are levels in A.

```
y ~ A*B + Error(C)
```

An experiment with two treatment factors, A and B, and error strata determined by factor C. For example, a split plot experiment, with whole plots (and hence also subplots), determined by factor C.

11.1.1 Contrasts

We need at least some idea how the model formulae specify the columns of the model matrix. This is easy if we have continuous variables, as each provides one column of the model matrix (and the intercept will provide a column of ones if included in the model).

What about a k-level factor `A`? The answer differs for unordered and ordered factors. For *unordered* factors $k-1$ columns are generated for the indicators of the second, ..., kth levels of the factor. (Thus the implicit parameterization is to contrast the response at each level with that at the first.) For *ordered* factors the $k-1$ columns are the orthogonal polynomials on $1, \ldots, k$, omitting the constant term.

Although the answer is already complicated, it is not the whole story. First, if the intercept is omitted in a model that contains a factor term, the first such term is encoded into k columns giving the indicators for all the levels. Second, the whole behavior can be changed by the `options` setting for `contrasts`. The default setting in R is

```
options(contrasts = c("contr.treatment", "contr.poly"))
```

The main reason for mentioning this is that R and S have different defaults for unordered factors, S using Helmert contrasts. So if you need to compare your results to those of a textbook or paper which used S-PLUS, you will need to set

```
options(contrasts = c("contr.helmert", "contr.poly"))
```

This is a deliberate difference, as treatment contrasts (R's default) are thought easier for newcomers to interpret.

We have still not finished, as the contrast scheme to be used can be set for each term in the model using the functions `contrasts` and `C`. And we have not yet considered interaction terms either: these generate the products of the columns introduced for their component terms.

Although the details are complicated, model formulae in R will normally generate the models that an expert statistician would expect, provided that marginality is preserved. For example, fitting a model with interaction but not the corresponding main effects will in general lead to surprising results, and is for experts only.

11.2 Linear models

The basic function for fitting ordinary multiple models is `lm()`, and a streamlined version of the call is as follows:

```
> fitted.model <- lm(formula, data = data.frame)
```

For example

```
> fm2 <- lm(y ~ x1 + x2, data = production)
```

would fit a multiple regression model of y on $x1$ and $x2$ (with implicit intercept term).

The important (but technically optional) parameter `data = production` specifies that any variables needed to construct the model should come first from the `production` *data frame. This is the case regardless of whether data frame* `production` *has been attached on the search path or not.*

By default, the fit is computed using ordinary least squares. If the optional argument `weights=w` is supplied, a weighted least squares fit is performed with the vector of weights *w*. It is also possible to restrict the fit to subsets of the data. See the entry for `lm()` in the R Reference Manual or help pages for further details of the fitting options.

11.3 Generic functions for extracting model information

The value of `lm()` is a fitted model object; technically a list of results of class `"lm"`. Information about the fitted model can then be displayed, extracted, plotted and so on by using generic functions that orient themselves to objects of class `"lm"`. These include

```
add1     coef        effects    kappa    predict   residuals
alias    deviance    family     labels   print     step
anova    drop1       formula    plot     proj      summary
```

A brief description of the most commonly used ones is given below:

`anova(object_1, object_2)`

Compare a submodel with an outer model and produce an analysis of variance table.

`coefficients(object)`
: Extract the regression coefficient (matrix).

  Short form: `coef(object)`.

`deviance(object)`
: Residual sum of squares, weighted if appropriate.

`formula(object)`
: Extract the model formula.

`plot(object)`
: Produce four plots, showing residuals, fitted values and some diagnostics.

`predict(object, newdata=data.frame)`
: The data frame supplied must have variables specified with the same labels as the original. The value is a vector or matrix of predicted values corresponding to the determining variable values in *data.frame*.

`print(object)`
: Print a concise version of the object. Most often used implicitly.

`residuals(object)`
: Extract the (matrix of) residuals, weighted as appropriate.

  Short form: `resid(object)`.

`step(object)`
: Select a suitable model by adding or dropping terms and preserving hierarchies. The model with the largest value of AIC (Akaike's An Information Criterion) discovered in the stepwise search is returned.

`summary(object)`
: Print a comprehensive summary of the results of the regression analysis.

11.4 Analysis of variance and model comparison

The model fitting function `aov(formula, data=data.frame)` fits an analysis of variance model. At the simplest level it operates in a very similar way to the function `lm()`, and most of the generic functions listed in the table in Section 11.3 [Generic functions for extracting model information], page 79 apply.

It should be noted that `aov()` also allows an analysis of models with multiple error strata such as split plot experiments, or balanced incomplete block designs with recovery of inter-block information. The model formula

```
response ~ mean.formula + Error(strata.formula)
```

specifies a multi-stratum experiment with error strata defined by the *strata.formula*. In the simplest case, *strata.formula* is simply a factor, when it defines a two strata experiment, namely between and within the levels of the factor.

For example, with all determining variables factors, a model formula such as that in

```
> fm <- aov(yield ~ v + n*p*k + Error(farms/blocks),
            data=farm.data)
```

would typically be used to describe an experiment with mean model v + n*p*k and three error strata, namely "between farms", "within farms, between blocks" and "within blocks".

11.4.1 ANOVA tables

Note also that the analysis of variance table (or tables) are for a sequence of fitted models. The sums of squares shown are the decrease in the residual sums of squares resulting from an inclusion of *that term* in the model at *that place* in the sequence. Hence only for orthogonal experiments will the order of inclusion be inconsequential.

For multistratum experiments, the procedure is first to project the response onto the error strata, again in sequence, and to fit the mean model to each projection. For further details, see Chambers & Hastie (1992).

A more flexible alternative to the default full ANOVA table is to compare two or more models directly using the anova() function:

```
> anova(fitted.model.1, fitted.model.2, ...)
```

The display is then an ANOVA table showing the differences between the fitted models when fitted in sequence. The fitted models being compared would usually be an hierarchical sequence, of course. This does not give different information to the default, but rather makes it easier to comprehend and control.

11.5 Updating fitted models

The update() function is largely a convenience function that allows a model to be fitted that differs from one previously fitted, usually by just a few additional or removed terms. Its form is

```
> new.model <- update(old.model, new.formula)
```

In the *new.formula* the special name consisting of a period '.', can be used to stand for "the corresponding part of the old model formula". For example,

```
> fm05 <- lm(y ~ x1 + x2 + x3 + x4 + x5,
             data = production)
> fm6  <- update(fm05, . ~ . + x6)
> smf6 <- update(fm6, sqrt(.) ~ .)
```

would fit a five variate multiple regression with variables (presumably) from the data frame `production`, then fit an additional model including a sixth regressor variable, and finally fit a variant on the model where the response had a square root transform applied.

Note especially that if the `data=` argument is specified on the original call to the model fitting function, this information is passed on through the fitted model object to `update()` and its allies.

The name '`.`' can also be used in other contexts, but with slightly different meaning. For example,

```
> fmfull <- lm(y ~ . , data = production)
```

would fit a model with response `y` and the regressor variables being *all other variables in the data frame* `production`.

Other functions for exploring incremental sequences of models are `add1()`, `drop1()` and `step()`. The names of these give a good clue to their purpose, but for full details see the R Reference Manual or help pages.

11.6 Generalized linear models

Generalized linear modeling is a development of linear models to accommodate both non-normal response distributions and transformations to linearity in a clean and straightforward way. A generalized linear model may be described in terms of the following sequence of assumptions:

- There is a response, y, of interest and stimulus variables x_1, x_2, ..., whose values influence the distribution of the response.
- The stimulus variables influence the distribution of y through *a single linear function, only*. This linear function is called the *linear predictor*, and is usually written

$$\eta = \beta_1 x_1 + \beta_2 x_2 + \cdots + \beta_p x_p,$$

  hence x_i has no influence on the distribution of y if and only if $\beta_i = 0$.
- The distribution of y is of the form

$$f_Y(y;\mu,\varphi) = \exp\left[\frac{A}{\varphi}\left\{y\lambda(\mu) - \gamma\left(\lambda(\mu)\right)\right\} + \tau(y,\varphi)\right]$$

  where φ is a *scale parameter* (possibly known), and is constant for all observations, A represents a prior weight, assumed known but possibly varying with the observations, and μ is the mean of y. So it is assumed that the distribution of y is determined by its mean and possibly a scale parameter as well.

- The mean, μ, is a smooth invertible function of the linear predictor:

$$\mu = m(\eta), \qquad \eta = m^{-1}(\mu) = \ell(\mu)$$

and this inverse function, $\ell()$, is called the *link function*.

These assumptions are loose enough to encompass a wide class of models useful in statistical practice, but tight enough to allow the development of a unified methodology of estimation and inference, at least approximately. The reader is referred to any of the current reference works on the subject for full details, such as McCullagh & Nelder (1989) or Dobson (1990).

11.6.1 Families

The class of generalized linear models handled by R includes *gaussian*, *binomial*, *poisson*, *inverse gaussian* and *gamma* response distributions and also *quasi-likelihood* models where the response distribution is not explicitly specified. In the latter case the *variance function* must be specified as a function of the mean, but in other cases this function is implied by the response distribution.

Each response distribution admits a variety of link functions to connect the mean with the linear predictor. Those automatically available are shown in the following table:

Family name	Link functions
`binomial`	`logit, probit, cloglog`
`gaussian`	`identity`
`Gamma`	`identity, inverse, log`
`inverse.gaussian`	`1/mu^2`
`poisson`	`identity, log, sqrt`
`quasi`	`logit, probit, cloglog, identity, inverse, log, 1/mu^2, sqrt`

The combination of a response distribution, a link function and various other pieces of information that are needed to carry out the modeling exercise is called the *family* of the generalized linear model.

11.6.2 The `glm()` function

Since the distribution of the response depends on the stimulus variables through a single linear function *only*, the same mechanism that was used for linear models can still be used to specify the linear part of a generalized model. Only the family has to be specified in a different way.

The R function to fit a generalized linear model is `glm()`, which uses the form

```
> fitted.model <-
   glm(formula, family=family.generator, data=data.frame)
```

The only new feature is the parameter *family.generator*, which is the instrument by which the family is described. It is the name of a function that generates a list of functions and expressions that together define and control the model and estimation process. Although this may seem a little complicated at first sight, its use is quite simple.

The names of the standard, supplied family generators are given under "Family Name" in the table in Section 11.6.1 [Families], page 83. Where there is a choice of links, the name of the link may also be supplied with the family name, in parentheses as a parameter. In the case of the `quasi` family, the variance function may also be specified in this way.

The examples below should make the process clear.

The `gaussian` family

A call such as

```
> fm <- glm(y ~ x1 + x2, family = gaussian, data = sales)
```

achieves the same result as

```
> fm <- lm(y ~ x1 + x2, data=sales)
```

using generalized linear model fitting (although less efficiently than a direct linear fit). Note how the gaussian family is not automatically provided with a choice of links, so no parameter is allowed. If a problem requires a gaussian family with a nonstandard link, this can usually be achieved through the `quasi` family, as we shall see later.

The `binomial` family

Consider a small, artificial example, from Silvey (1970).

On the Aegean island of Kalythos the male inhabitants suffer from a congenital eye disease, the effects of which become more marked with increasing age. Samples of islander males of various ages were tested for blindness and the results recorded. The data is shown below:

Age:	20	35	45	55	70
No. tested:	50	50	50	50	50
No. blind:	6	17	26	37	44

The problem we consider is to fit both logistic and probit models to this data, and to estimate for each model the LD50, that is the age at which the chance of blindness for a male inhabitant is 50%.

If y is the number of blind at age x and n the number tested, both models have the form

$$y \sim \mathrm{B}(n, F(\beta_0 + \beta_1 x))$$

where for the probit case, $F(z) = \Phi(z)$ is the standard normal distribution function, and in the logit case (the default), $F(z) = e^z/(1 + e^z)$. In both cases the LD50 is

$$\mathrm{LD50} = -\beta_0/\beta_1$$

that is, the point at which the argument of the distribution function is zero.

The first step is to set the data up as a data frame

```
> kalythos <- data.frame(x = c(20,35,45,55,70),
                         n = rep(50,5),
                         y = c(6,17,26,37,44))
```

To fit a binomial model using `glm()` there are two possibilities for the response:

- If the response is a *vector* it is assumed to hold *binary* data, and so must be a 0/1 vector.
- If the response is a *two column matrix* it is assumed that the first column holds the number of successes for the trial and the second holds the number of failures.

Here we need the second of these conventions, so we add a matrix to our data frame:

```
> kalythos$Ymat <- cbind(kalythos$y,
                         kalythos$n - kalythos$y)
```

To fit the models we use

```
> fmp <- glm(Ymat ~ x,
             family = binomial(link=probit),
             data = kalythos)
> fml <- glm(Ymat ~ x,
             family = binomial,
             data = kalythos)
```

Since the logit link is the default, the link parameter may be omitted on the second call. To see the results of each fit we could use

```
> summary(fmp)
> summary(fml)
```

Both models fit (all too) well. To find the LD50 estimate we can use a simple function:

```
> ld50 <- function(b) -b[1]/b[2]
> ldp <- ld50(coef(fmp)); ldl <- ld50(coef(fml));
> c(ldp, ldl)
```

The actual estimates from this data are 43.663 years and 43.601 years respectively.

Poisson models

With the Poisson family the default link is the `log`, and in practice the major use of this family is to fit surrogate Poisson log-linear models to frequency data, whose actual distribution is often multinomial. This is a large and important subject we will not discuss further here. It even forms a major part of the use of non-gaussian generalized models overall.

Occasionally, genuinely Poisson data arises in practice and in the past it was often analyzed as gaussian data after either a log or a square-root transformation. As a graceful alternative to the latter, a Poisson generalized linear model may be fitted as in the following example:

```
> fmod <- glm(y ~ A + B + x,
              family = poisson(link=sqrt),
              data = worm.counts)
```

Quasi-likelihood models

For all families, the variance of the response will depend on the mean and will have the scale parameter as a multiplier. The form of dependence of the variance on the mean is a characteristic of the response distribution; for example, for the poisson distribution $\mathrm{Var}[y] = \mu$.

For quasi-likelihood estimation and inference the precise response distribution is not specified, but rather only a link function and the form of the variance function as it depends on the mean. Since quasi-likelihood estimation uses formally identical techniques to those for the gaussian distribution, this family provides a way of fitting gaussian models with non-standard link functions or variance functions.

For example, consider fitting the non-linear regression

$$y = \frac{\theta_1 z_1}{z_2 - \theta_2} + e$$

which may be written alternatively as

$$y = \frac{1}{\beta_1 x_1 + \beta_2 x_2} + e$$

where $x_1 = z_2/z_1$, $x_2 = -1/x_1$, $\beta_1 = 1/\theta_1$ and $\beta_2 = \theta_2/\theta_1$. Supposing a suitable data frame to be set up we could fit this non-linear regression as:

```
> nlfit <-
    glm(y ~ x1 + x2 - 1,
        family = quasi(link=inverse, variance=constant),
        data = biochem)
```

The reader is referred to the R Reference Manual and help pages for further information, as needed.

11.7 Nonlinear least squares and maximum likelihood models

Certain forms of nonlinear model can be fitted by Generalized Linear Models using `glm()`. But in the majority of cases we have to approach the nonlinear curve fitting problem as one of nonlinear optimization. R's nonlinear optimization routine is `nlm()`, which takes the place of S-PLUS's `ms()` and `nlmin()`. We seek the parameter values that minimize some index of lack-of-fit, and `nlm()` does that by trying out various parameter values iteratively. Unlike linear regression, there is no guarantee that the procedure will converge on satisfactory estimates. The `nlm()` function requires initial guesses about what parameter values to try, and convergence depends critically upon the quality of the initial guesses.

11.7.1 Least squares

One way to fit a nonlinear model is by minimizing the sum of the squared errors (SSE) or residuals. This method makes sense if the observed errors could have plausibly arisen from a normal distribution.

Here is an example from Bates & Watts (1988), page 51. The data are:

```
> x <- c(0.02, 0.02, 0.06, 0.06, 0.11, 0.11, 0.22, 0.22,
         0.56, 0.56, 1.10, 1.10)
> y <- c(76, 47, 97, 107, 123, 139, 159, 152, 191, 201,
         207, 200)
```

And the model to be fitted is:

```
> fn <- function(p) sum((y - (p[1] * x)/(p[2] + x))^2)
```

In order to do the fit we need initial estimates of the parameters. One way to find sensible starting values is to plot the data, guess some parameter values, and superimpose the model curve using those values. We could do better, but starting values of 200 and .1 seem adequate:

```
> plot(x, y)
> xfit <- seq(.02, 1.1, .05)
> yfit <- 200 * xfit/(0.1 + xfit)
> lines(spline(xfit, yfit))
```

Now we do the fit:

```
> out <- nlm(fn, p = c(200, 0.1), hessian = TRUE)
```

After the fitting, `out$minimum` is the sum of squared errors, and `out$estimate` are the least squares estimates of the parameters.

```
> out
$minimum
[1] 1195.449

$estimate
[1] 212.68384222   0.06412146

$gradient
[1] -0.0001535012  0.0934206673

$hessian
            [,1]        [,2]
[1,]     11.94725   -7661.319
[2,] -7661.31875 8039421.153

$code
[1] 3

$iterations
[1] 26
```

To obtain the approximate standard errors (SE) of the estimates we do:

```
> sqrt(diag(2*out$minimum/(length(y) - 2)
         * solve(out$hessian)))
```

The 2 in the line above represents the number of parameters. A 95% confidence interval would be the parameter estimate $\pm$ 1.96 SE. We can superimpose the least squares fit on a new plot:

```
> plot(x, y)
> xfit <- seq(.02, 1.1, .05)
> yfit <- out$estimate[1] * xfit/(out$estimate[2] + xfit)
> lines(spline(xfit, yfit))
```

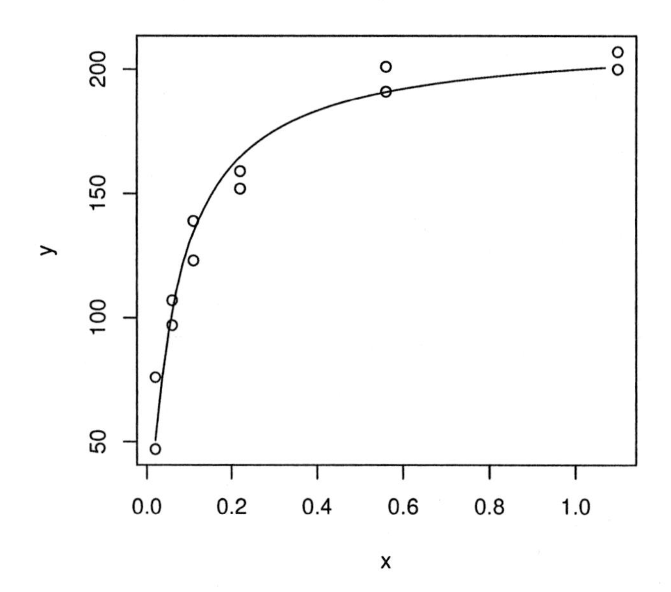

The standard package **stats** provides much more extensive functionality for fitting non-linear models by least squares. The model we have just fitted is the Michaelis-Menten model, so we can use the following commands to obtain the same result. With older versions of R, it may be necessary to load the nls fitting function first with the command library(nls).

```
> df <- data.frame(x=x, y=y)
> fit <- nls(y ~ SSmicmen(x, Vm, K), df)
> fit
Nonlinear regression model
  model:  y ~ SSmicmen(x, Vm, K)
   data:  df
          Vm             K
212.68370711   0.06412123
 residual sum-of-squares:  1195.449
> summary(fit)

Formula: y ~ SSmicmen(x, Vm, K)

Parameters:
    Estimate Std. Error t value Pr(>|t|)
Vm 2.127e+02  6.947e+00  30.615 3.24e-11
K  6.412e-02  8.281e-03   7.743 1.57e-05

Residual standard error: 10.93 on 10 degrees of freedom
```

```
Correlation of Parameter Estimates:
      Vm
K 0.7651
```

11.7.2 Maximum likelihood

Maximum likelihood is a method of nonlinear model fitting that applies even if the errors are not normal. The method finds the parameter values which maximize the log likelihood, or equivalently which minimize the negative log-likelihood. Here is an example from Dobson (1990), pp. 108–111. This example fits a logistic model to dose-response data, which clearly could also be fit by `glm()`. The data are:

```
> x <- c(1.6907, 1.7242, 1.7552, 1.7842, 1.8113,
         1.8369, 1.8610, 1.8839)
> y <- c( 6, 13, 18, 28, 52, 53, 61, 60)
> n <- c(59, 60, 62, 56, 63, 59, 62, 60)
```

The negative log-likelihood to minimize is:

```
> fn <- function(p)
   sum( - (y*(p[1]+p[2]*x) - n*log(1+exp(p[1]+p[2]*x))
          + log(choose(n, y)) ))
```

We pick sensible starting values and do the fit:

```
> out <- nlm(fn, p = c(-50,20), hessian = TRUE)
```

After the fitting, `out$minimum` is the negative log-likelihood, and `out$estimate` are the maximum likelihood estimates of the parameters. To obtain the approximate SEs of the estimates we do:

```
> sqrt(diag(solve(out$hessian)))
```

A 95% confidence interval would be the parameter estimate $\pm$ 1.96 SE.

11.8 Some non-standard models

We conclude this chapter with just a brief mention of some of the other features available in R for special regression and data analysis problems.

- **Mixed models.** The user-contributed **nlme** package provides functions `lme()` and `nlme()` for linear and non-linear mixed-effects models, that is linear and non-linear regressions in which some of the coefficients correspond to random effects. These functions make heavy use of formulae to specify the models.
- **Local approximating regressions.** The `loess()` function fits a nonparametric regression by using a locally weighted regression. Such

regressions are useful for highlighting a trend in messy data or for data reduction to give some insight into a large data set.

Function `loess` is in the standard package **stats**, together with code for projection pursuit regression.

- **Robust regression.** There are several functions available for fitting regression models in a way resistant to the influence of extreme outliers in the data. Function `lqs` in the recommended package **MASS** provides state-of-art algorithms for highly-resistant fits. Less resistant but statistically more efficient methods are available in packages, such as the function `rlm` in package **MASS**.
- **Additive models.** This technique aims to construct a regression function from smooth additive functions of the determining variables, usually one for each determining variable. Functions `avas` and `ace` in package **acepack** and functions `bruto` and `mars` in package **mda** provide some examples of these techniques in user-contributed packages to R.
- **Tree-based models.** Rather than seek an explicit global linear model for prediction or interpretation, tree-based models seek to bifurcate the data, recursively, at critical points of the determining variables in order to partition the data ultimately into groups that are as homogeneous as possible within, and as heterogeneous as possible between. The results often lead to insights that other data analysis methods tend not to yield.

  Models are again specified in the ordinary linear model form. The model fitting function is `tree()`, but many other generic functions such as `plot()` and `text()` are well adapted to displaying the results of a tree-based model fit in a graphical way.

  Tree models are available in R *via* the user-contributed packages **rpart** and **tree**.

12 Graphical procedures

Graphics are an important and extremely versatile feature of the R environment. It is possible to display a wide variety of statistical plots, and also to build entirely new types of graph.

The graphics facilities can be used in both interactive and batch modes, but in most cases, interactive use is more productive. Interactive use is also easy because at startup time R initiates a graphics *device driver* which opens a special *graphics window* for the display of interactive graphics. Although this is done automatically, it is useful to know that the command used is `X11()` under UNIX and `windows()` under Windows.

Once the device driver is running, R plotting commands can be used to produce a variety of graphs and to create entirely new kinds of display.

Plotting commands are divided into three basic groups:

- **High-level** plotting functions create a new plot on the graphics device, possibly with axes, labels, titles and so on.
- **Low-level** plotting functions add more information to an existing plot, such as extra points, lines and labels.
- **Interactive** graphics functions allow you interactively add information to, or extract information from, an existing plot using a pointing device such as a mouse.

In addition, R maintains a list of *graphical parameters* which can be manipulated to customize your plots.

This manual only describes what are known as 'base' graphics. A separate graphics sub-system in package **grid** coexists with base—it is more powerful but harder to use. There is a recommended package **lattice** which builds on **grid** and provides ways to produce multi-panel plots akin to those in the *Trellis* system in S.

12.1 High-level plotting commands

High-level plotting functions are designed to generate a complete plot of the data passed as arguments to the function. Where appropriate, axes, labels and titles are automatically generated (unless you request otherwise). High-level plotting commands always start a new plot, erasing the current plot if necessary.

12.1.1 The `plot()` function

One of the most frequently used plotting commands in R is the `plot()` function. This is a *generic* function: the type of plot produced is dependent on the type or *class* of the first argument.

`plot(x, y)`
`plot(pts)`
: If *x* and *y* are vectors, `plot(x, y)` produces a scatterplot of *y* against *x*. The same effect can be produced by supplying one argument (*pts*) as either a list containing two elements *x* and *y* or a two-column matrix.

`plot(x)`
: If *x* is a time series, this produces a time-series plot. If *x* is a numeric vector, it produces a plot of the values in the vector against their index in the vector. If *x* is a complex vector, it produces a plot of imaginary versus real parts of the vector elements.

`plot(f)`
`plot(f, y)`
: *f* is a factor object, *y* is a numeric vector. The first form generates a bar plot of *f*; the second form produces boxplots of *y* for each level of *f*.

`plot(df)`
`plot(~ expr)`
`plot(y ~ expr)`
: *df* is a data frame, *y* is any object, *expr* is a list of object names separated by '+' (e.g., `a + b + c`). The first two forms produce distributional plots of the variables in a data frame (first form) or of a number of named objects (second form). The third form plots *y* against every object named in *expr*.

12.1.2 Displaying multivariate data

R provides two very useful functions for representing multivariate data. If `X` is a numeric matrix or data frame, the command

```
> pairs(X)
```

produces a pairwise scatterplot matrix of the variables defined by the columns of `X`, that is, every column of `X` is plotted against every other column of `X` and the resulting $n(n-1)$ plots are arranged in a matrix with plot scales constant over the rows and columns of the matrix.

When three or four variables are involved a *coplot* may be more enlightening. If `a` and `b` are numeric vectors and `c` is a numeric vector or factor object (all of the same length), then the command

```
> coplot(a ~ b | c)
```

produces a number of scatterplots of a against b for given values of c. If c is a factor, this simply means that a is plotted against b for every level of c. When c is numeric, it is divided into a number of *conditioning intervals* and for each interval a is plotted against b for values of c within the interval. The number and position of intervals can be controlled with the `given.values=` argument to `coplot()`—the function `co.intervals()` is useful for selecting intervals here. You can also use two *given* variables with a command like

```
> coplot(a ~ b | c + d)
```

which produces scatterplots of a against b for every joint conditioning interval of c and d.

The `coplot()` and `pairs()` function both take an argument `panel=` which can be used to customize the type of plot which appears in each panel. The default is `points()` to produce a scatterplot, but by supplying some other low-level graphics function of two vectors x and y as the value of `panel=` you can produce any type of plot you wish. An example panel function useful for coplots is `panel.smooth()`.

12.1.3 Display graphics

Other high-level graphics functions produce different types of plots. Some examples are:

`qqnorm(x)`
`qqline(x)`
`qqplot(x, y)`
: Distribution-comparison plots. The first form plots the numeric vector x against the expected Normal order scores (a normal scores plot) and the second adds a straight line to such a plot by drawing a line through the distribution and data quartiles. The third form plots the quantiles of x against those of y to compare their respective distributions.

`hist(x)`
`hist(x, nclass=`*n*`)`
`hist(x, breaks=`*b*`, ...)`
: Produces a histogram of the numeric vector x. A sensible number of classes is usually chosen, but a recommendation can be given with the `nclass=` argument. Alternatively, the breakpoints can be specified exactly with the `breaks=` argument. If the `probability=TRUE` argument is given, the bars represent relative frequencies instead of counts.

`dotchart(x, ...)`

Constructs a dotchart of the data in x. In a dotchart, the y-axis gives a labelling of the data in x and the x-axis gives its value. For example, it allows easy visual selection of all data entries with values lying in specified ranges.

`image(x, y, z, ...)`
`contour(x, y, z, ...)`
`persp(x, y, z, ...)`

Plots of three variables. The `image` plot draws a grid of rectangles using different colors to represent the value of `z`, the `contour` plot draws contour lines to represent the value of `z`, and the `persp` plot draws a 3D surface.

12.1.4 Arguments to high-level plotting functions

There are a number of arguments which may be passed to high-level graphics functions, as follows:

`add=TRUE`

Forces the function to act as a low-level graphics function, superimposing the plot on the current plot (some functions only).

`axes=FALSE`

Suppresses generation of axes—useful for adding your own custom axes with the `axis()` function. The default, `axes=TRUE`, means include axes.

`log="x"`
`log="y"`
`log="xy"`

Causes the x, y or both axes to be logarithmic. This will work for many, but not all, types of plot.

`type=`

The `type=` argument controls the type of plot produced, as follows:

`type="p"`

Plot individual points (the default)

`type="l"`

Plot lines

`type="b"`

Plot points connected by lines ("both" points and lines)

`type="o"`

Plot points overlaid by lines

`type="h"`
: Plot vertical lines from points to the zero axis (*high-density*)

`type="s"`
`type="S"`
: Step-function plots. In the first form, the top of the vertical defines the point; in the second, the bottom.

`type="n"`
: No plotting at all. However axes are still drawn (by default) and the coordinate system is set up according to the data. Ideal for creating plots with subsequent low-level graphics functions.

`xlab=`*`string`*
`ylab=`*`string`*
: Axis labels for the x and y axes. Use these arguments to change the default labels, usually the names of the objects used in the call to the high-level plotting function.

`main=`*`string`*
: Figure title, placed at the top of the plot in a large font.

`sub=`*`string`*
: Sub-title, placed just below the x-axis in a smaller font.

12.2 Low-level plotting commands

Sometimes the high-level plotting functions don't produce exactly the kind of plot you desire. In this case, low-level plotting commands can be used to add extra information (such as points, lines or text) to the current plot.

Some of the more useful low-level plotting functions are:

`points(x, y)`
`lines(x, y)`
: Add points or connected lines to the current plot. A `type=` argument can also be passed to these functions (and defaults to `"p"` for `points()` and `"l"` for `lines()`).

`text(x, y, labels, ...)`
: Add text to a plot at points given by `x, y`. Normally `labels` is an integer or character vector in which case `labels[i]` is plotted at the point `(x[i], y[i])`. The default is `1:length(x)`.

: The optional argument pos= specifies the position of the text. Values of '1', '2', '3' and '4', indicate positions below, to the left of, above and to the right of the specified coordinates respectively.

Note: This function is often used to display a set of labelled points with the sequence

```
> plot(x, y, type="n"); text(x, y, names)
```

The graphics parameter `type="n"` suppresses the points but sets up the axes, and the `text()` function supplies special characters, as specified by the character vector `names` for the points.

`abline(a, b)`
`abline(h=`*y*`)`
`abline(v=`*x*`)`
`abline(`*lm.obj*`)`

Adds a line of slope `b` and intercept `a` to the current plot. `h=`*y* may be used to specify y-coordinates for the heights of horizontal lines to go across a plot, and `v=`*x* similarly for the x-coordinates for vertical lines. Also *lm.obj* may be a list with a `coefficients` component of length 2 (such as the result of model-fitting functions), which are taken as an intercept and slope, in that order.

`polygon(x, y, ...)`

Draws a polygon defined by the ordered vertices in (`x`, `y`). The polygon can also be shaded with hatch lines, or filled in if the graphics device allows the filling of figures—see the R Reference Manual or help for details.

`legend(x, y, legend, ...)`

Adds a legend to the current plot at the specified position. Plotting characters, line styles, colors etc., are identified with the labels in the character vector `legend`. At least one other argument *v* (a vector the same length as `legend`) with the corresponding values of the plotting unit must also be given, as follows:

`legend( , fill=`*v*`)`
Colors for filled boxes

`legend( , col=`*v*`)`
Colors in which points or lines will be drawn

`legend( , lty=`*v*`)`
Line styles

`legend( , lwd=`*v*`)`
Line widths

`legend( , pch=`*v*`)`
Plotting characters (character vector)

`title(main, sub)`

Adds a title `main` to the top of the current plot in a large font and (optionally) a sub-title `sub` at the bottom in a smaller font.

`axis(side, ...)`
: Adds an axis to the current plot on the side given by the first argument (1 to 4, counting clockwise from the bottom). Other arguments control the positioning of the axis within or beside the plot, and tick positions and labels. Useful for adding custom axes after calling `plot()` with the `axes=FALSE` argument.

Low-level plotting functions usually require some positioning information (e.g., x and y coordinates) to determine where to place the new plot elements. Coordinates are given in terms of *user coordinates* which are defined by the previous high-level graphics command and are chosen based on the supplied data.

Where `x` and `y` arguments are required, it is also sufficient to supply a single argument being a list with elements named `x` and `y`. Similarly, a matrix with two columns is also valid input. In this way functions such as `locator()` (see below) may be used to specify positions on a plot interactively.

12.2.1 Mathematical annotation

In some cases, it is useful to add mathematical symbols and formulae to a plot. This can be achieved in R by specifying an *expression* rather than a character string in any one of `text`, `mtext`, `axis`, or `title`. For example, the following code draws the formula for the Binomial probability function:

```
> text(x, y,
     expression(paste(bgroup("(", atop(n, x), ")"),
                     p^x, q^{n-x})))
```

More information, including a full listing of the features available can be found in the R Reference Manual entry for '`plotmath`', or obtained within R using the commands:

```
> help(plotmath)
> example(plotmath)
```

12.2.2 Hershey vector fonts

The standard distribution of R includes a set of public-domain vector fonts, known as the Hershey fonts[(1)]. It is possible to specify Hershey vector fonts for rendering text when using the `text` and `contour` functions. There are three reasons for using the Hershey fonts:

---

(1) The fonts were originally developed by Dr. A. V. Hershey, and published in coordinate form by the U.S. National Bureau of Standards in 1976.

- Hershey fonts can produce better output, especially on a computer screen, for rotated and/or small text.
- Hershey fonts provide some symbols that may not be available in the standard fonts. In particular, there are zodiac signs, cartographic symbols and astronomical symbols.
- Hershey fonts provide Cyrillic and Japanese (Kana and Kanji) characters.

More information, including tables of Hershey characters can be found in the R Reference Manual, or from within R using the commands:

```
> help(Hershey)
> example(Hershey)
> help(Japanese)
> example(Japanese)
```

12.3 Interacting with graphics

R also provides functions which allow users to extract or add information to a plot using a mouse. The simplest of these is the `locator()` function:

`locator(n, type)`
: Waits for the user to select locations on the current plot using the left mouse button. This continues until `n` points have been selected, or another mouse button is pressed. The `type` argument allows for plotting at the selected points and has the same effect as for high-level graphics commands; the default is no plotting. `locator()` returns the locations of the points selected as a list with two components `x` and `y`.

The `locator()` function is often called with no arguments, and in this case is particularly useful for interactively selecting positions for graphic elements, such as legends or labels (where it is difficult to calculate an appropriate position in advance). For example, to place some informative text near an outlying point, the command

```
> text(locator(1), "Outlier", pos=1)
```

may be useful. `locator()` will still work if the current device does not support a mouse; in this case the user will be prompted for x and y coordinates. When called with no arguments, the `locator()` function has a large default value for `n` of 512.

`identify(x, y, labels)`
: Allow the user to highlight any of the points defined by `x` and `y` (using the left mouse button) by plotting the corresponding component of `labels` nearby (or the index number of the point if `labels`

is absent). Returns the indices of the selected points when another button is pressed.

Sometimes we want to identify particular *points* on a plot, rather than their positions. For example, we may wish the user to select some observation of interest from a graphical display and then manipulate that observation in some way. Given a number of (x, y) coordinates in two numeric vectors x and y, we could use the identify() function as follows:

```
> plot(x, y)
> identify(x, y)
```

The identify() function performs no plotting itself, but simply allows the user to move the mouse pointer and click the left mouse button near a point. The point nearest the mouse pointer will be highlighted with its index number (that is, its position in the (x, y) vectors) plotted nearby. Alternatively, you could use some informative string (such as a case name) as a highlight by using the labels argument to identify(), or disable highlighting altogether with the plot=FALSE argument. When the process is terminated (see above), identify() returns the indices of the selected points; you can use these indices to extract the selected points from the original vectors x and y.

12.4 Using graphics parameters

When creating graphics, particularly for presentation or publication purposes, the default output from R sometimes needs to be customized. You can change almost every aspect of the display using *graphics parameters*. R maintains a list of a large number of graphics parameters which control features such as line style, colors, figure arrangement and text justification among many others. Every graphics parameter has a name (such as 'col', which controls colors) and a value (a color number, for example).

A separate list of graphics parameters is maintained for each active device, and each device has a default set of parameters when initialized. Graphics parameters can be set in two ways: either permanently, affecting all graphics functions which access the current device; or temporarily, affecting only a single graphics function call.

12.4.1 Permanent changes: The par() function

The par() function is used to access and modify the list of graphics parameters for the current graphics device.

par()

Without arguments, returns a list of all graphics parameters and their values for the current device.

`par(c("col", "lty"))`
: With a character vector argument, returns only the named graphics parameters (again, as a list).

`par(col=4, lty=2)`
: With named arguments (or a single list argument), sets the values of the named graphics parameters, and returns the original values of the parameters as a list.

Setting graphics parameters with the `par()` function changes the value of the parameters *permanently*, in the sense that all future calls to graphics functions (on the current device) will be affected by the new value. You can think of setting graphics parameters in this way as setting "default" values for the parameters, which will be used by all graphics functions unless an alternative value is given.

Note that calls to `par()` *always* affect the global values of graphics parameters, even when `par()` is called from within a function. This is often undesirable behavior—usually we want to set some graphics parameters, do some plotting, and then restore the original values so as not to affect the user's R session. You can restore the initial values by saving the result of `par()` when making changes, and restoring the initial values when plotting is complete.

```
> oldpar <- par(col=4, lty=2)
  ... plotting commands ...
> par(oldpar)
```

12.4.2 Temporary changes: Arguments to graphics functions

Graphics parameters may also be passed to (almost) any graphics function as named arguments. This has the same effect as passing the arguments to the `par()` function, except that the changes only last for the duration of the function call. For example:

```
> plot(x, y, pch="+")
```

produces a scatterplot using a plus sign as the plotting character, without changing the default plotting character for future plots.

12.5 Graphics parameters list

The following sections detail many of the commonly-used graphical parameters. The entry for the `par()` function in the R Reference Manual or help documentation provides a more concise summary.

Graphics parameters will be presented in the following form:

`name=value`
: A description of the parameter's effect. *name* is the name of the parameter, that is, the argument name to use in calls to `par()` or a graphics function. *value* is a typical value you might use when setting the parameter.

12.5.1 Graphical elements

R plots are made up of points, lines, text and polygons (filled regions). The following graphical parameters control how these *graphical elements* are drawn:

`pch=4`
: Symbol to be used for plotting points, an integer between 0 and 18 inclusive. Each value of `pch` corresponds to a specialized plotting symbol—these include squares, circles, triangles, crosses and stars. The default varies with graphics drivers, but it is usually '∘'.

The first ten symbols are,

0 — open square
1 — open circle
2 — open triangle
3 — cross
4 — diagonal cross
5 — diamond
6 — inverted triangle
7 — crossed square
8 — asterisk
9 — crossed diamond

To see the complete range of symbols, use the command

```
> plot(0, 0, type="n")
> legend(0, 1, as.character(0:18), pch=0:18)
```

`pch="A"`
: The plotting symbol `pch` can also be a character. When characters are used as plotting symbols, the character may appear fractionally above or below the appropriate position, depending on the font and character shape.

`lty=2`
: Line types. Alternative line styles are not supported on all graphics devices (and vary on those that do) but line type 1 is always a solid line, and line types 2 and onwards are dotted or dashed lines, or some combination of both.

`lwd=2`
: Line widths. Desired width of lines, in multiples of the "standard" line width. Affects axis lines as well as lines drawn with `lines()`, etc.

`col=2`
: Colors to be used for points, lines, text, filled regions and images. Each of these graphic elements has a list of possible colors, and the value of this parameter is an index to that list. Obviously, this parameter applies only to devices supporting colors.

`font=2`
: An integer which specifies which font to use for text. If possible, device drivers arrange so that 1 corresponds to plain text, 2 to bold face, 3 to italic and 4 to bold italic.

`font.axis`
`font.lab`
`font.main`
`font.sub`
: The font to be used for axis annotation, x and y labels, main and sub-titles, respectively.

`adj=-0.1`
: Justification of text relative to the plotting position. `0` means left justify, `1` means right justify and `0.5` means to center horizontally about the plotting position. The actual value is the proportion of text that appears to the left of the plotting position, so a value of `-0.1` leaves a gap of 10% of the text width between the text and the plotting position.

`cex=1.5`
: Character expansion. The value is the desired size of text characters (including plotting characters) relative to the default text size.

12.5.2 Axes and tick marks

Many of R's high-level plots have axes, and you can construct axes yourself with the low-level `axis()` graphics function. Axes have three main components: the *axis line* (line style controlled by the `lty` graphics parameter), the *tick marks* (which mark off unit divisions along the axis line) and the *tick labels* (which mark the units). These components can be customized with the following graphics parameters:

`lab=c(5, 7, 12)`
: The first two numbers are the desired number of tick intervals on the x and y axes respectively. The third number is the desired

length of axis labels, in characters (including the decimal point). Choosing a too-small value for this parameter may result in all tick labels being rounded to the same number!

`las=1`
: Orientation of axis labels. 0 means always parallel to axis, 1 means always horizontal, and 2 means always perpendicular to the axis.

`mgp=c(3, 1, 0)`
: Positions of axis components. The first component is the distance from the axis label to the axis position, in text lines. The second component is the distance to the tick labels, and the final component is the distance from the axis position to the axis line (usually zero). Positive numbers measure outside the plot region, negative numbers inside.

`tck=0.01`
: Length of tick marks, as a fraction of the smaller of the width or height of the plotting region. When `tck` is less than 0.5 the tick marks on the x and y axes are forced to be the same size, otherwise `tck` is interpreted as a fraction of the relevant side. A value of 1 gives grid lines. Negative values give tick marks outside the plotting region. Use `tck=0.01` and `mgp=c(1,-1.5,0)` for internal tick marks.

`xaxs="s"`
`yaxs="d"`
: Axis styles for the x and y axes, respectively. With styles `"s"` (standard) and `"e"` (extended) the smallest and largest tick marks always lie outside the range of the data. Extended axes may be widened slightly if any points are very near the edge. This style of axis can sometimes leave large blank gaps near the edges. With styles `"i"` (internal) and `"r"` (the default) tick marks always fall within the range of the data, however style `"r"` leaves a small amount of space at the edges.

  Setting this parameter to `"d"` (direct axis) *locks in* the current axis and uses it for all future plots (or until the parameter is set to one of the other values above, at least). This is useful for generating a series of fixed-scale plots.

12.5.3 Figure margins

A single plot in R is known as a `figure` and comprises a *plot region* surrounded by margins (possibly containing axis labels, titles, etc.) and (usually) bounded by the axes themselves.

A typical figure is

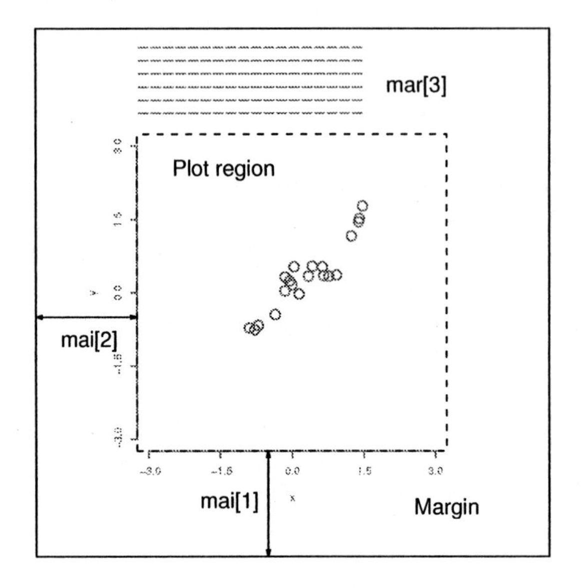

Graphics parameters controlling figure layout include:

`mai=c(1, 0.5, 0.5, 0)`
: Widths of the bottom, left, top and right margins, respectively, measured in inches.

`mar=c(4, 2, 2, 1)`
: Similar to `mai`, except the measurement unit is text lines.

`mar` and `mai` are equivalent in the sense that setting one changes the value of the other. The default values chosen for this parameter are often too large; the right-hand margin is rarely needed, and neither is the top margin if no title is being used. The bottom and left margins must be large enough to accommodate the axis and tick labels. Furthermore, the default is chosen without regard to the size of the device surface: for example, using the `postscript()` driver with the `height=4` argument will result in a plot which is about 50% margin unless `mar` or `mai` are set explicitly. When multiple figures are in use (see below) the margins are reduced by half, however this may not be enough when many figures share the same page.

12.5.4 Multiple figure environment

R allows you to create an n by m array of figures on a single page. Each figure has its own margins, and the array of figures is optionally surrounded by an *outer margin*, as shown in the following figure.

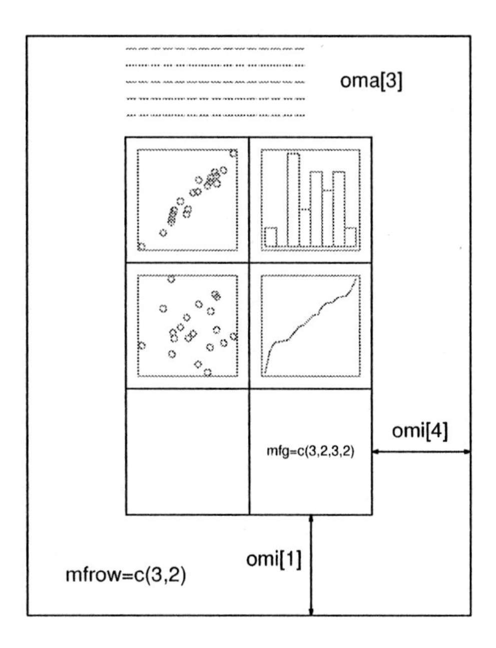

The graphical parameters relating to multiple figures are as follows:

`mfcol=c(3, 2)`
`mfrow=c(2, 4)`

Set the size of a multiple figure array. The first value is the number of rows; the second is the number of columns. The only difference between these two parameters is that setting `mfcol` causes figures to be filled by column; `mfrow` fills by rows.

The layout in the Figure could have been created by setting `mfrow=c(3,2)`; the figure shows the page after four plots have been drawn.

Setting either of these can reduce the base size of symbols and text (controlled by `par("cex")` and the pointsize of the device). In a layout with exactly two rows and columns the base size is reduced by a factor of 0.83: if there are three or more of either rows or columns, the reduction factor is 0.66.

`mfg=c(2, 2, 3, 2)`
: Position of the current figure in a multiple figure environment. The first two numbers are the row and column of the current figure; the last two are the number of rows and columns in the multiple figure array. Set this parameter to jump between figures in the array. You can even use different values for the last two numbers than the *true* values for unequally-sized figures on the same page.

`fig=c(4, 9, 1, 4)/10`
: Position of the current figure on the page. Values are the positions of the left, right, bottom and top edges respectively, as a percentage of the page measured from the bottom left corner. The example value would be for a figure in the bottom right of the page. Set this parameter for arbitrary positioning of figures within a page. If you want to add a figure to a current page, use `new=TRUE` as well (unlike S).

`oma=c(2, 0, 3, 0)`
`omi=c(0, 0, 0.8, 0)`
: Size of outer margins. Like `mar` and `mai`, the first measures in text lines and the second in inches, starting with the bottom margin and working clockwise.

Outer margins are particularly useful for page-wise titles, etc. Text can be added to the outer margins with the `mtext()` function with argument `outer=TRUE`. There are no outer margins by default, however, so you must create them explicitly using `oma` or `omi`.

More complicated arrangements of multiple figures can be produced with the `split.screen()` and `layout()` functions.

12.6 Device drivers

It is possible to generate graphics on almost any type of display or printing device with R, by starting the appropriate *device driver*. The device driver converts graphical instructions from R ("draw a line," for example) into a form that the particular device can understand.

Device drivers are started by calling a device driver function. There is one such function for every device driver: see the entry for `Devices` in the R Reference Manual, or type `help(Devices)` at the R prompt for a list of them all. For example, issuing the command

```
> postscript()
```

causes all future graphics output to be sent to the default system printer in PostScript format. Some commonly-used device drivers are:

`X11()`
: For use with the X11 window system on Unix-like systems

`windows()`
: For use on Windows

`quartz()`
: For use on MacOS X

`postscript()`
: For printing on PostScript printers, or creating PostScript graphics files.

`pdf()`
: Produces a PDF file, which can also be included in other PDF files.

`png()`
: Produces a bitmap PNG file.

`jpeg()`
: Produces a bitmap JPEG file, best used for `image` plots.

R can be compiled without the PNG and JPEG devices, so these are not always available in every R installation: see the entries for `png()` and `jpeg()` in the R Reference Manual or help pages for details.

When you have finished with a device, be sure to terminate the device driver by issuing the command

```
> dev.off()
```

This shuts down the device cleanly; for example, in the case of hardcopy devices it ensures that any remaining pages are completed and sent to the printer.

12.6.1 PostScript diagrams for typeset documents

By passing the `file` argument to the `postscript()` device driver function, you may store the graphics in PostScript format in a file of your choice. The plot will be in landscape orientation unless the `horizontal=FALSE` argument is given, and you can control the size of the graphic with the `width` and `height` arguments (the plot will be scaled as appropriate to fit these dimensions). For example, the command

```
> postscript("file.ps", horizontal=FALSE, height=5,
             pointsize=10)
```

will produce a file containing PostScript code for a figure five inches high, perhaps for inclusion in a document. It is important to note that if the file named in the command already exists, it will be overwritten. This is the case even if the file was only created earlier in the same R session.

Many uses of PostScript output will involve incorporating the figure in another document. This works best when *encapsulated* PostScript (EPS) is selected: R produces output in the EPS format when the `onefile=FALSE` argument is supplied. This unusual notation stems from S-compatibility: it really means that the output will be a single page (which is part of the EPS specification). Thus to produce a plot for inclusion use something like

```
> postscript("plot1.eps",
             horizontal=FALSE, onefile=FALSE,
             height=8, width=6, pointsize=10)
```

12.6.2 Multiple graphics devices

In advanced usage of R it is often necessary to have several graphics devices open at the same time. Of course, only one graphics device can accept graphics commands at any given time, and this is known as the *current device.* When multiple devices are open, they form a numbered sequence with names giving the kind of device at any position.

Each new call to a device driver function, such as `X11()`, `postscript()` or `png()`, opens a new graphics device, thus extending the device list by one. This new device becomes the current device, to which graphics output will then be sent.

The main commands for working with multiple devices are as follows:

`dev.list()`
: Returns the number and name of all active devices. The device at position 1 on the list is always the *null device* which does not accept graphics commands at all.

`dev.next()`
`dev.prev()`
: Returns the number and name of the graphics device following, or prior to the current device, respectively.

`dev.set(which=`*k*`)`
: Can be used to change the current graphics device to the one at position *k* of the device list. Returns the number and label of the device.

`dev.off(`*k*`)`
: Terminate the graphics device at point *k* of the device list. For some devices, such as `postscript` devices, this will either print the file immediately or correctly complete the file for later printing, depending on how the device was initiated.

`dev.copy(device, ..., which=k)`
`dev.print(device, ..., which=k)`

Make a copy of the device k. Here `device` is a device function, such as `postscript`, with extra arguments, if needed, specified by '...'. `dev.print` is similar, but the copied device is immediately closed, so that end actions, such as printing hardcopies, are immediately performed.

`graphics.off()`

Terminate all graphics devices on the list, except the null device.

13 Packages

All R functions and datasets are stored in *packages*. Only when a package is loaded are its contents available. This is done both for efficiency (the full list would take more memory and would take longer to search than a subset), and to aid package developers, who are protected from name clashes with other code. The process of developing packages is described in the document *Writing R Extensions* supplied with R. Here we will describe them from a user's point of view.

To see which packages are installed at your site, issue the command

```
> library()
```

with no arguments. To load a particular package (e.g., the **boot** package containing functions from Davison & Hinkley (1997)), use a command like

```
> library(boot)
```

Users connected to the Internet can use the `CRAN.packages()` function[1] to automatically update and install packages.

To see which packages are currently loaded, use

```
> search()
```

to display the search list. A few packages may be loaded but not available on the search list (see Section 13.3 [Namespaces], page 115).

To see a list of all available help topics in an installed package, use

```
> help(package = "name")
```

A complete list of the user-level objects in a package can be displayed using the command

```
> ls("package:name")
```

For example, `ls("package:base")` displays all the base package commands. Alternatively, if you have a web-browser available, you can use

```
> help.start()
```

to start the HTML help system, and then navigate to the package listing in the `Reference` section.

13.1 Standard packages

The standard (or *base*) packages are considered part of the R source code. They contain the basic functions that allow R to work, and the standard statistical and graphical functions that are described in this tutorial. They

---

[1] Also available through the `Packages` menu in the Windows and RAqua GUIs

are automatically available in any R installation, and are documented in the R Reference Manual.

13.2 Contributed packages and CRAN

There are hundreds of contributed packages for R, written by many different authors. Some of these packages implement specialized statistical methods, others give access to data or hardware, and others are designed to complement textbooks. Some (the *recommended* packages) are distributed with every binary distribution of R. Most are available for download from CRAN, the Comprehensive R Archive Network[2], and other repositories such as Bioconductor[3]. The *R FAQ* contains a list that was current at the time of release, but the collection of available packages changes frequently.

The recommended packages included with R are:

`boot` Bootstrap R (S-Plus) Functions (Canty)
: Functions and datasets for bootstrapping from the book "Bootstrap Methods and Their Applications" by A. C. Davison and D. V. Hinkley (1997, CUP).

`cluster` Functions for clustering (by Rousseeuw et al.)
: Functions for cluster analysis, originally from Peter Rousseeuw, Anja Struyf and Mia Hubert.

`foreign` Read data stored by other programs
: Functions for reading and writing data stored by statistical packages such as Minitab, S, SAS, SPSS, Stata, ...

`KernSmooth` Functions for kernel smoothing for Wand & Jones (1995)
: Functions for kernel smoothing (and density estimation) corresponding to the book: Wand, M.P. and Jones, M.C. (1995) "Kernel Smoothing".

`lattice` Lattice Graphics
: Implementation of Trellis Graphics.

`mgcv` Multiple smoothing parameter estimation and GAMs by GCV
: Routines for generalized additive models (GAMs) and other generalized ridge regression with multiple smoothing parameter selection by generalized cross validation (GCV) or Un-Biased Risk Estimator (UBRE). Includes a `gam()` function.

---

[2] `http://cran.r-project.org/` and its mirrors

[3] `http://www.bioconductor.org/`

`nlme` Linear and nonlinear mixed effects models
: Fit and compare Gaussian linear and nonlinear mixed-effects models.

`rpart` Recursive Partitioning
: Recursive partitioning and regression trees.

`survival` Survival analysis, including penalised likelihood.
: Survival analysis: descriptive statistics, two-sample tests, parametric accelerated failure models, Cox model. Delayed entry (truncation) allowed for all models; interval censoring for parametric models.

`VR` (this provides `MASS`, `class`, `nnet` and `spatial`)
: Functions and datasets to support Venables and Ripley, "Modern Applied Statistics with S" (4th edition).

13.3 Namespaces

Packages can have *namespaces*, and currently all of the base and most of the recommended packages do. Namespaces do three things: they allow the package writer to hide functions and data that are meant only for internal use, they prevent functions from breaking when a user (or other package writer) picks a name that clashes with one in the package, and they provide a way to refer to an object within a particular package.

For example, `t()` is the transpose function in R, but users might define their own function named `t`. Namespaces prevent the user's definition from taking precedence, and breaking every function that tries to transpose a matrix.

There are two operators that work with namespaces. The double-colon operator `::` selects definitions from a particular namespace. In the example above, the transpose function will always be available as `base::t`, because it is defined in the `base` package. Only functions that are exported from the package can be retrieved in this way.

The triple-colon operator `:::` may be seen in a few places in R code: it acts like the double-colon operator but also allows access to hidden objects. Users are more likely to use the `getAnywhere()` function, which searches multiple packages.

Packages are often inter-dependent, and loading one may cause others to be automatically loaded. The colon operators described above will also cause automatic loading of the associated package. When packages with namespaces are loaded automatically they are not added to the search list.

Appendix A A sample session

The following session is intended to introduce to you some features of the R environment by using them. Many features of the system will be unfamiliar and puzzling at first, but this puzzlement will soon disappear.

Login, start your windowing system.

`$ R` Start R as appropriate for your platform.

The R program begins, with a banner.

(Within R, the prompt on the left hand side ('>') will not be shown to avoid confusion.)

```
help.start()
```
Start the HTML interface to on-line help (using a web browser available at your machine). You should briefly explore the features of this facility with the mouse.

Iconify the help window and move on to the next part.

```
x <- rnorm(50)
y <- rnorm(x)
```
Generate two pseudo-random normal vectors of x- and y-coordinates.

```
plot(x, y)
```
Plot the points in the plane. A graphics window will appear automatically.

`ls()` See which R objects are now in the R workspace.

```
rm(x, y)
```
Remove objects no longer needed. (Clean up).

```
x <- 1:20
```
Make $x = (1, 2, \ldots, 20)$.

```
w <- 1 + sqrt(x)/2
```
A 'weight' vector of standard deviations.

```
dummy <- data.frame(x = x, y = x + rnorm(x)*w)
dummy
```
Make a *data frame* of two columns, x and y, and look at it.

```
fm <- lm(y ~ x, data=dummy)
summary(fm)
```
Fit a simple linear regression of y on x, and look at the analysis.

```
fm1 <- lm(y ~ x, data=dummy, weight=1/w^2)
summary(fm1)
```

Since we know the standard deviations, we can do a weighted regression.

```
attach(dummy)
```

Make the columns in the data frame visible as variables.

```
lrf <- lowess(x, y)
```

Make a nonparametric local regression function.

```
plot(x, y)
```

Standard point plot.

```
lines(x, lrf$y)
```

Add in the local regression.

```
abline(0, 1, lty=3)
```

The true regression line: (intercept 0, slope 1).

```
abline(coef(fm))
```

Unweighted regression line.

```
abline(coef(fm1), col="red")
```

Weighted regression line.

```
detach()
```

Remove data frame from the search path.

```
plot(fitted(fm), resid(fm),
    xlab="Fitted values",
    ylab="Residuals",
    main="Residuals vs Fitted")
```

A standard regression diagnostic plot to check for heteroscedasticity. Can you see it?

```
qqnorm(resid(fm), main="Residuals Rankit Plot")
```

A normal scores plot to check for skewness, kurtosis and outliers. (Not very useful here.)

```
rm(fm, fm1, lrf, x, dummy)
```

Clean up again.

The next section will look at data from the classical experiment of Michaelson and Morley to measure the speed of light.

You will need a copy of the the file 'morley.tab' in your working directory. This file can be found in the 'base/data' subdirectory of the default R library tree, or downloaded from the publishers website[1].

---

[1] http://www.network-theory.co.uk/download/R/

`file.show("morley.tab")`
: Optional. Look at the file.

  If you do not have this file, use the command `data(morley)` and replace the next command by `mm <- morley`.

```
mm <- read.table("morley.tab")
mm
```
Read in the Michaelson and Morley data as a data frame, and look at it. There are five experiments (column `Expt`) and each has 20 runs (column `Run`) and `sl` is the recorded speed of light, suitably coded.

```
mm$Expt <- factor(mm$Expt)
mm$Run <- factor(mm$Run)
```
Change `Expt` and `Run` into factors.

```
attach(mm)
```
Make the data frame visible at position 2 (the default).

```
plot(Expt, Speed, main="Speed of Light Data", xlab="Experiment No.")
```
Compare the five experiments with simple boxplots.

```
fm <- aov(Speed ~ Run + Expt, data=mm)
summary(fm)
```
Analyze as a randomized block, with 'runs' and 'experiments' as factors.

```
fm0 <- update(fm, . ~ . - Run)
anova(fm0, fm)
```
Fit the sub-model omitting 'runs', and compare using a formal analysis of variance.

```
detach()
rm(fm, fm0)
```
Clean up before moving on.

We now look at some more graphical features: contour and image plots.

```
x <- seq(-pi, pi, len=50)
y <- x
```
x is a vector of 50 equally spaced values in $-\pi \le x \le \pi$. y is the same.

```
f <- outer(x, y, function(x, y) cos(y)/(1 + x^2))
```
f is a square matrix, with rows and columns indexed by x and y respectively, of values of the function $\cos(y)/(1 + x^2)$.

```
oldpar <- par(no.readonly = TRUE)
par(pty="s")
```
Save the plotting parameters and set the plotting region to "square".

```
contour(x, y, f)
contour(x, y, f, nlevels=15, add=TRUE)
```
Make a contour map of f; add in more lines for more detail.

```
fa <- (f-t(f))/2
```
`fa` is the "asymmetric part" of f (the matrix transpose of `f` is given by `t(f)`).

```
contour(x, y, fa, nlevels=15)
```
Make a contour plot, ...

```
par(oldpar)
```
... and restore the old graphics parameters.

```
image(x, y, f)
image(x, y, fa)
```
Make some high density image plots (of which you can get hardcopies if you wish), ...

```
objects(); rm(x, y, f, fa)
```
... and clean up before moving on.

R can also do complex arithmetic:

```
th <- seq(-pi, pi, len=100)
z <- exp(1i*th)
```
`1i` is used for the complex number i.

```
par(pty="s")
plot(z, type="l")
```
Plotting complex arguments means plot imaginary versus real parts. This should be a circle.

```
w <- rnorm(100) + rnorm(100)*1i
```
Suppose we want to sample points within the unit circle. One method would be to take complex numbers with standard normal real and imaginary parts ...

```
w <- ifelse(Mod(w) > 1, 1/w, w)
```
... and to map any outside the circle onto their reciprocal.

```
plot(w, xlim=c(-1,1), ylim=c(-1,1), pch="+", xlab="x",
ylab="y")
lines(z)
```
All points are inside the unit circle, but the distribution is not uniform.

```
w <- sqrt(runif(100))*exp(2*pi*runif(100)*1i)
plot(w, xlim=c(-1,1), ylim=c(-1,1), pch="+", xlab="x",
ylab="y")
lines(z)
```

The second method uses the uniform distribution. The points should now look more evenly spaced over the disc.

```
rm(th, w, z)
```

Clean up again.

`q()` Quit the R program. You will be asked if you want to save the R workspace, and for an exploratory session like this, you probably do not want to save it.

Appendix B Invoking R

B.1 Invoking R from the command line

When working in UNIX or at a command line in Windows, the command 'R' is used to start the main R program in the form

R [*options*] [<*infile*] [>*outfile*],

Input and output can be redirected in the usual way, using '<' and '>' in the UNIX shell.

Many of the command-line options control what happens at the beginning and at the end of an R session. The full startup mechanism is given below (see also the topic 'Startup' in the R Reference Manual or help pages for more information).

Note that under UNIX you do need to ensure that if the environment variable TMPDIR is set that it points to a valid place to create temporary files and directories.

- Unless '--no-environ' was given, R searches for user and site files to process for setting environment variables. The name of the site file is the one pointed to by the environment variable R_ENVIRON; if this is unset, '$R_HOME/etc/Renviron.site' is used instead (if it exists). The user files searched for are '.Renviron' in the current directory or in the user's home directory (in that order). These files should contain lines of the form '*name*=*value*'. (See help(Startup) for a precise description.) Variables you might want to set include R_PAPERSIZE (the default paper size), R_PRINTCMD (the default print command) and R_LIBS (specifies the list of R library trees searched for add-on packages).
- Then R searches for the site-wide startup profile unless the command line option '--no-site-file' was given. The name of this file is taken from the value of the R_PROFILE environment variable. If that variable is unset, the default '$R_HOME/etc/Rprofile.site' is used if this exists.
- Then, unless '--no-init-file' was given, R searches for a file called '.Rprofile' in the current directory or in the user's home directory (in that order) and sources it.
- It also loads a saved image from '.RData' if there is one (unless '--no-restore' or '--no-restore-data' was specified).

- Finally, if a function `.First` exists, it is executed. This function (as well as `.Last` which is executed at the end of the R session) can be defined in the appropriate startup profiles, or reside in '`.RData`'.

In addition, there are options for controlling the memory available to the R process (see the R Reference Manual or help for the topic '`Memory`' for more information). Users will not normally need to use these unless they are trying to limit the amount of memory used by R.

R accepts the following command-line options:

'`--help`'
'`-h`' Print short help message to standard output and exit.

'`--version`'
Print version information to standard output and exit.

'`RHOME`'
Print the path to the R "home directory" to standard output and exit. Apart from the front-end shell script and the man page, the standard R installation puts everything (executables, packages, etc.) into this directory.

'`--save`'
'`--no-save`'
Control whether data sets should be saved or not at the end of the R session. If neither is given in an interactive session, the user is asked for the desired behavior when ending the session with *q()*; in batch mode, one of these must be specified.

'`--no-environ`'
Do not read any user file to set environment variables.

'`--no-site-file`'
Do not read the site-wide profile at startup.

'`--no-init-file`'
Do not read the user's profile at startup.

'`--restore`'
'`--no-restore`'
'`--no-restore-data`'
Control whether saved images should be restored at startup or not. The default is to restore. The name of the image file is '`.RData`', in the directory where R was started. The option ('`--no-restore`' implies all the specific '`--no-restore-*`' options.

'`--no-restore-history`'
Control whether the history file should be restored at startup or not. The default is to restore. The history file is normally

'.Rhistory' in the directory where R was started, but can be set by the environment variable R_HISTFILE.

'--vanilla'
: Combine '--no-save', '--no-environ' '--no-site-file', '--no-init-file' and '--no-restore'.

'--no-readline'
: (UNIX only) Turn off command-line editing via **readline**. This is useful when running R from within Emacs using the ESS package ("Emacs Speaks Statistics"). See Appendix C [The command line editor], page 129, for more information.

'--ess'
: (Windows only) Set Rterm up for use by R-inferior-mode in ESS.

'--min-vsize=*N*'
'--max-vsize=*N*'
: Specify the minimum or maximum amount of memory used for variable size objects by setting the "vector heap" size to *N* bytes. Here, *N* must either be an integer or an integer ending with 'G', 'M', 'K', or 'k', meaning 'Giga' (2^30), 'Mega' (2^20), (computer) 'Kilo' (2^10), or regular 'kilo' (1000).

'--min-nsize=*N*'
'--max-nsize=*N*'
: Specify the amount of memory used for fixed size objects by setting the number of "cons cells" to *N*. See the previous option for details on *N*. A cons cell takes 28 bytes on a 32-bit machine, and usually 56 bytes on a 64-bit machine.

'--max-ppsize=*N*'
: Specify the maximum size of the pointer protection stack as *N* locations. This defaults to 10000, but can be increased to allow large and complicated calculations to be done. Currently the maximum value accepted is 100000.

'--max-mem-size=*N*'
: (Windows only) Specify a limit for the amount of memory to be used both for R objects and working areas. This is set by default to the smaller of 1024Mb and the amount of physical RAM in the machine, and must be at least 16Mb.

'--quiet'
'--silent'
'-q' Do not print out the initial copyright and welcome messages.

'--slave'

Make R run as quietly as possible. This option is intended to support programs which use R to compute results for them.

'--verbose'

Print more information about progress, and in particular set R's option `verbose` to `TRUE`. R code uses this option to control the printing of diagnostic messages.

'--debugger=*name*'
'-d *name*'

(UNIX only) Run R through debugger *name*. Note that in this case, further command line options are disregarded, and should instead be given when starting the R executable from inside the debugger.

'--gui=*type*'
'-g *type*'

(UNIX only) Use *type* as graphical user interface (note that this also includes interactive graphics). Currently, possible values for *type* are 'X11' (the default), 'gnome' provided that GNOME support is available, and 'none'.

'--args'

This flag does nothing except cause the rest of the command line to be skipped: this can be useful to retrieve values from it with `commandArgs()`.

The command `R CMD` acts as a wrapper to various R tools (e.g., for processing files in R documentation format, or manipulating add-on packages) which are useful in conjunction with R, but not intended to be called "directly". The general form is

```
R CMD command args
```

where *command* is the name of the tool and *args* the arguments passed on to it.

Currently, the following tools are available.

`BATCH`

Run R in batch mode.

`COMPILE`

(UNIX only) Compile files for use with R.

`SHLIB`

Build shared library for dynamic loading.

`INSTALL`

Install add-on packages.

`REMOVE`
: Remove add-on packages.

`build`
: Build add-on packages.

`check`
: Check add-on packages.

`LINK`
: (UNIX only) Front-end for creating executable programs.

`Rprof`
: Post-process R profiling files.

`Rdconv`
: Convert Rd format to various other formats, including HTML, Nroff, LaTeX, plain text, and S documentation format.

`Rd2dvi`
: Convert Rd format to DVI/PDF.

`Rd2txt`
: Convert Rd format to text.

`Sd2Rd`
: Convert S documentation to Rd format.

`config`
: (UNIX only) Obtain configuration information.

The first five tools (i.e., `BATCH`, `COMPILE`, `SHLIB`, `INSTALL`, and `REMOVE`) can also be invoked "directly" without the `CMD` option, i.e., in the form `R` *`command args`*.

Use the command

```
R CMD command --help
```

to obtain usage information for each of the tools accessible via the `R CMD` interface.

B.2 Invoking R under Windows

There are two ways to run R under Windows. Within a terminal window, the methods described in the previous section may be used, by invoking `R.exe` or, more directly, `Rterm.exe`. These versions are principally intended for batch use. For interactive use, there is a console-based graphical user interface, `Rgui.exe`.

The startup procedure under Windows is very similar to that under UNIX, but references to the 'home directory' need to be clarified, as this is not always defined on Windows. If the environment variable `R_USER` is defined, that gives the home directory. If not, and the environment variable HOME is defined, that gives the home directory instead. Otherwise, if the environment variables HOMEDRIVE and HOMEPATH are set (and they normally are under Windows NT/2000/XP) they define the home directory. Failing all those, the home directory is taken to be the starting directory.

Environment variables can be supplied as '*name=value*' pairs at the end of the command line.

The following additional command-line options are available when invoking RGui.exe.

'--mdi'
'--sdi'
'--no-mdi'
: Control whether Rgui will operate as an MDI program (the default, with multiple child windows within one main window) or an SDI application (with multiple top-level windows for the console, graphics and pager).

'--debug'
: Enable the "Break to debugger" menu item in Rgui, and trigger a break to the debugger during command line processing.

In Windows you may also specify your own *.bat or *.exe file with R CMD instead of one of the built-in commands. It will be run with the following environment variables set appropriately: R_HOME, R_VERSION, R_CMD, R_OSTYPE, PATH, PERL5LIB, and TEXINPUTS. For example, if you already have latex.exe on your path, then

```
R CMD latex.exe mydoc
```

will run LaTeX on mydoc.tex, with the path to R's share/texmf macros added to TEXINPUTS.

B.3 Invoking R under Mac OS X

There are two ways to run R under Mac OS X. The first is within a Terminal.app window by invoking R, where the methods described in the previous sections apply. There is also console-based GUI (R.app) that is installed in the Applications folder on your system by default. It is a standard double-clickable Mac OS X application.

The startup procedure under Mac OS X is very similar to that under UNIX. The 'home directory' is the one inside the R.framework. But the startup and current working directory is set as the user's home directory unless a different startup directory is given in the Preferences window accessible from within the GUI.

Appendix C The command line editor

C.1 Preliminaries

On Unix systems, R will use the GNU **readline** library when possible to allow the interactive recall, editing and re-submission of prior commands[1]. The readline interface can be disabled using the startup option '`--no-readline`' if necessary.[2].

Windows versions of R have somewhat simpler command-line editing: see '`Console`' under the '`Help`' menu of the GUI, and the file '`README.Rterm`' for command-line editing under `Rterm.exe`.

When using R with **readline** capabilities, the functions described below are available. Many of these use either Control or Meta characters. Control characters, such as *`Control-n`*, are obtained by holding the control key ⟨CTRL⟩ down while you press the ⟨n⟩ key, and are written as *`C-n`* below. Meta characters, such as *`Meta-b`*, are typed by holding down the "meta" key ⟨META⟩ and pressing ⟨b⟩, and written as *`M-b`* in the following. On many systems the ⟨ALT⟩ key is used as the meta key. If your terminal does not have a ⟨META⟩ key, you can still type Meta characters using two-character sequences starting with *ESC*. Thus, to enter *`M-b`*, you could type ⟨ESC⟩⟨b⟩. The *ESC* character sequences are also allowed on terminals with real Meta keys. Note that case is significant for Meta characters.

C.2 Editing actions

The R program keeps a history of the commands you type, including the error lines. The commands in your history may be recalled, changed if necessary, and re-submitted as new commands. In Emacs-style command-line editing, any straight typing you do while in this editing phase causes the characters to be inserted in the command you are editing, displacing any characters to the right of the cursor. In *vi* mode, character insertion mode is started by *`M-i`* or *`M-a`*, characters are typed and insertion mode is finished by typing a further ⟨ESC⟩.

Pressing the ⟨RET⟩ command at any time causes the command to be re-submitted.

---

(1) Readline is not used in the GNOME interface under UNIX

(2) This is needed for use with ESS, the 'Emacs Speaks Statistics' package; see the URL `http://ess.stat.wisc.edu/`

Other editing actions are summarized in the following table.

C.3 Command line editor summary

Command recall and vertical motion

`C-p` Go to the previous command (backwards in the history).

`C-n` Go to the next command (forwards in the history).

`C-r text`
Find the last command with the *text* string in it.

On most terminals, you can also use the up and down arrow keys instead of `C-p` and `C-n`, respectively.

Horizontal motion of the cursor

`C-a` Go to the beginning of the command.

`C-e` Go to the end of the line.

`M-b` Go back one word.

`M-f` Go forward one word.

`C-b` Go back one character.

`C-f` Go forward one character.

On most terminals, you can also use the left and right arrow keys instead of `C-b` and `C-f`, respectively.

Editing and re-submission

`text`
Insert *text* at the cursor.

`C-f text`
Append *text* after the cursor.

⟨BACKSPACE⟩ *(or* ⟨DEL⟩*)*
Delete the previous character (left of the cursor).

`C-d` Delete the character under the cursor.

`M-d` Delete the rest of the word under the cursor, and "save" it.

`C-k` Delete from the cursor to the end of the command line, and "save" it.

`C-y` Insert (yank) the last "saved" text here.

`C-t` Transpose the character under the cursor with the next.

`M-l` Change the rest of the word to lower case.

`M-c` Change the rest of the word to upper case.

⟨RET⟩
Re-submit the command to R.

The final ⟨RET⟩ terminates the command-line editing sequence.

Appendix D References

D. M. Bates and D. G. Watts (1988), *Nonlinear Regression Analysis and Its Applications.* John Wiley & Sons, New York.

Richard A. Becker, John M. Chambers and Allan R. Wilks (1988), *The New S Language.* Chapman & Hall, New York. This book is often called the "*Blue Book*".

John M. Chambers and Trevor J. Hastie eds. (1992), *Statistical Models in S.* Chapman & Hall, New York. This is also called the "*White Book*".

A. C. Davison and D. V. Hinkley (1997), *Bootstrap Methods and Their Applications,* Cambridge University Press.

Annette J. Dobson (1990), *An Introduction to Generalized Linear Models,* Chapman and Hall, London.

Peter McCullagh and John A. Nelder (1989), *Generalized Linear Models.* Second edition, Chapman and Hall, London.

John A. Rice (1995), *Mathematical Statistics and Data Analysis.* Second edition. Duxbury Press, Belmont, CA.

S. D. Silvey (1970), *Statistical Inference.* Penguin, London.

Further Reading

The R Reference Manual (2 volumes) is the most comprehensive published documentation available for R, with over 1,400 pages of detailed information on its commands and functions.

The two volumes explain the use of every standard R command, with numerous examples, implementation notes and references to the statistical literature. Each volume includes a comprehensive index of entries, listed by keyword and topic.

The full bibliographic details for the printed edition of the R Reference Manual are:

- *The R Reference Manual—Base Package (Volume 1)* by the R Development Core Team (ISBN 0-9546120-0-0, hardback) $69.95 (£39.95) Covers all the fundamental R commands and functions.
- *The R Reference Manual—Base Package (Volume 2)* by the R Development Core Team (ISBN 0-9546120-1-9, hardback) $69.95 (£39.95). Covers graphics, mathematics, distributions, models, time-series and datasets.

The text of these manuals has been written by the developers of R, and is used to provide the online help facility in R itself. For each set of manuals sold (volumes 1 & 2), $10 is donated to the R Foundation by the publisher.

Printed copies of the R Reference Manual are available for order from all major bookstores and library suppliers worldwide, including Amazon and Barnes & Noble. Further information about the two volumes can be found on the web at `http://www.network-theory.co.uk/R/`

Brian Gough (Publisher)
Network Theory Ltd.

Other Books from the Publisher

Network Theory Ltd publishes books about free software under free documentation licenses. In addition to *An Introduction to R* and the R Reference Manuals, our current catalogue includes the following titles:

- *An Introduction to GCC* by Brian Gough, foreword by Richard Stallman (ISBN 0-9541617-9-3) $19.95 (£12.95)
- *GNU Octave Manual* by John W. Eaton (ISBN 0-9541617-2-6) $29.99 (£19.99)
- *GNU Scientific Library Reference Manual—Second Edition* by M. Galassi, et al (ISBN 0-9541617-3-4) $39.99 (£24.99)
- *Comparing and Merging Files with GNU diff and patch* by David MacKenzie, Paul Eggert, and Richard Stallman (ISBN 0-9541617-5-0) $19.95 (£12.95)
- *Version Management with CVS* by Per Cederqvist et al. (ISBN 0-9541617-1-8) $29.95 (£19.95)
- *GNU Bash Reference Manual* by Chet Ramey and Brian Fox (ISBN 0-9541617-7-7) $29.95 (£19.95)
- *An Introduction to Python* by Guido van Rossum and Fred L. Drake, Jr. (ISBN 0-9541617-6-9) $19.95 (£12.95)
- *Python Language Reference Manual* by Guido van Rossum and Fred L. Drake, Jr. (ISBN 0-9541617-8-5) $19.95 (£12.95)

All titles are available for order from bookstores worldwide. Sales of the manuals fund the development of more free software and documentation. Bulk order discounts are available for academic, corporate and government purchases.

For further details, related resources and ordering information, visit our website at `http://www.network-theory.co.uk/` or contact us by email at `sales@network-theory.co.uk`.

Index

E

F

G

H

I

J

K

L

M

N

O

P

Q

R

S

T

U

V

W

X

Printed in the United States
46721LVS00006B/127-165

9 780954 161743